THE CHEMISTRY
BETWEEN US

CURRENT

THE CHEMISTRY BETWEEN US

LOVE, SEX, AND THE
SCIENCE OF ATTRACTION

Larry Young, PhD,
and
Brian Alexander

CURRENT

CURRENT
Published by the Penguin Group
Penguin Group (USA) Inc., 375 Hudson Street, New York, New York 10014, U.S.A. • Penguin
Group (Canada), 90 Eglinton Avenue East, Suite 700, Toronto, Ontario, Canada M4P 2Y3
(a division of Pearson Penguin Canada Inc.) • Penguin Books Ltd, 80 Strand, London
WC2R 0RL, England • Penguin Ireland, 25 St. Stephen's Green, Dublin 2, Ireland (a division of
Penguin Books Ltd) • Penguin Books Australia Ltd, 250 Camberwell Road, Camberwell,
Victoria 3124, Australia (a division of Pearson Australia Group Pty Ltd) • Penguin Books India
Pvt Ltd, 11 Community Centre, Panchsheel Park, New Delhi – 110 017, India • Penguin Group
(NZ), 67 Apollo Drive, Rosedale, Auckland 0632, New Zealand (a division of Pearson New
Zealand Ltd) • Penguin Books (South Africa) (Pty) Ltd, 24 Sturdee Avenue, Rosebank,
Johannesburg 2196, South Africa

Penguin Books Ltd, Registered Offices:
80 Strand, London WC2R 0RL, England

First published in 2012 by Current,
a member of Penguin Group (USA) Inc.

10   9   8   7   6   5   4   3   2   1

Illustrations by Biqing Li.

LIBRARY OF CONGRESS CATALOGING IN PUBLICATION DATA
Young, Larry, Ph.D.
   The chemistry between us : love, sex, and the science of attraction / Larry Young and Brian
Alexander.
      p. cm.
   Includes bibliographical references and index.
   ISBN 978-1-59184-513-3
   1. Love—Psychological aspects.   2. Love—Physiological aspects.   3. Sexual attraction.
4. Sex (Psychology)   I. Alexander, Brian, 1977–   II. Title.
   BF575.L8Y68 2012
   612.8—dc23   2012019129

Printed in the United States of America
Set in Bulmer MT Std
Designed by Alissa Amell

ALWAYS LEARNING   PEARSON

To our families, where love lives

CONTENTS

Men are still led by instinct before they are regulated by knowledge.

—THEODORE DREISER, *SISTER CARRIE*

INTRODUCTION

The notion of love as "mystery" is so ingrained in the human imagination, it may be our oldest cliché. Plato called it "irrational desire." Cole Porter seemed to speak for most of us when he threw up his hands and, in a kind of musical sigh of resignation, asked, "What is this thing called love?" In his classic song, a person describes being content, even if leading a "humdrum" life, until love mysteriously "flew in," changed everything, and made "a fool of me."

At some time or other all of us are gobsmacked by the dramatic way our behavior can change when love flies in. And our desire for sex can seem boundless. We crave it so much, we'll lay out good money just to be reminded of it, to the financial benefit of Hugh Hefner, Jimmy Choo, and the economy of Las Vegas. The combination of erotic desire and the love it leads to may be the most powerful force on earth. Some people will kill for love. We'll marry somebody with children and happily accept the job of parenting their kids, though we had no intention of living with children of any kind when we were single. We'll change religions or join a religion for the first time. We'll leave the warm breezes of Miami and move to the frostbitten climes of Minnesota. We'll do any number of things we never imagined doing, think thoughts we never imagined thinking, adopt ways of living we never imagined adopting—all under the sway of love.

And if love goes bad, we'll wonder, like Porter's once-contented protagonist, what went wrong and how we could have been so foolish.

How does this happen? How can two complete strangers come to the conclusion that it would not only be pleasant to share their lives but that they *must* share them? How can a man say he loves his wife and yet have sex with another woman? What makes us stay in relationships even after the romance fades? How is it possible to fall in love with the "wrong" person? How do people come to have a "type"? How does love begin? What drives mothers to care for their babies? What accounts for the gender of the people at whom we aim our affection? What does it even mean to say one is female or male? Where does that idea reside? How is it formed?

When Larry began doctoral studies in neuroscience, within the zoology department of the University of Texas, he wasn't really thinking about trying to answer such questions. Then he started working with an unusual species of lizard. (We'll explain later why these lizards are unusual.) Lizards wouldn't seem to have much to tell us about human love, but Larry began to make the connection in his own mind when he discovered he was able to completely and exquisitely control the sexual behavior of these animals by administering just one chemical. A single molecule, active in the brain, created fundamental changes in mating behavior. For Larry personally, and for his scientific career, it was a profound moment. He wasn't the first to realize that one chemical or another had such power: as you'll see, earlier generations of scientists had paved the way. But as Larry researched their work, made discoveries of his own, and as he and others advanced the knowledge of social neuroscience—the study of how we behave in relation to others—a picture of the brain mechanisms underlying the mysteries that had befuddled so many for so long began to form in his mind. This book is an attempt to interpret that picture.

Throughout time, storytellers like Plato and Porter have performed much hand waving when trying to explain love, so daring to barge in where they've already trod may seem like a fool's errand. But we've teamed up to try, because new science is proving that Larry's instinct back in graduate school was correct: desire, love, and the bonds between people aren't so mysterious after all. Love doesn't really fly in and out. The complex behaviors surrounding these emotions are driven by a few molecules in our brains. It's these molecules, acting on defined neural circuits, that so powerfully influence some of the biggest, most life-changing decisions we'll ever make.

We tend to think of the miasma of symbols, incantations, and, ultimately, behaviors surrounding love as mysterious because we seem to have so little mastery over them. Yet we also like to believe that we're not driven by our most basic instincts, that being human insulates us from the force of passion. Humans, after all, have very large and complicated brain cortices—the frontal lobes. We've taken great comfort in these seats of reason, congratulating ourselves on having risen above our unthinking, impulse-driven animal cousins through long evolutionary advancement.

The Stanford University physician and neuroscientist Josef Parvizi calls this attitude "corticocentric bias." In fact, the brain is made up of a number of structures that respond to a galaxy of neurochemicals. Contrary to common wisdom, no one region of the brain is "higher" or "lower" than any other. Behavior doesn't always develop as part of a stepwise hierarchical process; it's more of a parallel operation. This doesn't mean we are completely at the mercy of our irrational drives, and we will not argue so in this book. Reason can indeed help us apply the brakes. But we shouldn't underestimate the power of those drives, either. The brain circuits of desire and love have such enormous influence, they routinely overrule our rational selves, making our behavior subject to the force of evolutionary drives. As Parvizi has written, in the nineteenth century "humans were considered to be strictly different from animals because of their voluntary inhibition of instinctual desires by virtue of rationality and pure reason. However, the times have changed. We have recently begun to acknowledge the biological bases of our core human values, such as empathy, sense of fairness, and culture in other animals."

You're going to read a lot about animals, as well as people, in this book. There are a couple of reasons why. Animals have a lot to tell us about our own sexual behaviors and social bonding—about human love. There's a tendency to use trite statements like "animals aren't people" as a way to dismiss the relevance to humans of animal behavior studies. True enough: Animals aren't people. But animals, even animals we think of as primitive, are under the influence of the same neurochemicals as people when it comes to mating and reproduction. Those chemicals trigger behaviors in the animals just as they do in people. These animal behaviors often have close human analogues, and the common neurochemicals, and neural circuits, have been evolutionarily conserved. Sometimes, in the human context, they've been repurposed, or tweaked, but they're there nonetheless, and they drive us to action.

You've probably seen television programs featuring functional magnetic resonance imaging (fMRI) and other such technologies used in experiments where a human brain is stimulated by music, or math problems, or NFL football and then depicted in gee-whiz, TV-friendly colors as this or that reacting brain region lights up in reds or greens. These experiments are valuable, and you'll read about some of them here. But fMRI and other imaging technologies aren't the only, or even the most powerful, tools available for teasing out the "hows" of behavior. They're used so often, and so enthusiastically, because there are few other ethical ways to peer inside the functioning human brain. Unfortunately, the results of such tests can be more suggestive than definitive. On the other hand, new techniques that allow scientists to manipulate the brain circuits of living animals are being employed to see how such manipulations change behavior, what chemicals are involved, and what brain events result. Such animal experiments, sometimes later corroborated with human imaging studies, have helped researchers understand the development of emotions like fear and anxiety. These findings have led to the creation of drugs used to treat humans for phobias and post-traumatic stress disorder.

Some may argue that sex and love are so enormously complex—so very mysterious—that relying on animals to teach us about our own sexual and romantic behaviors is a pretty big stretch. We fully expect such criticism. As you'll read, however, some animals—like the unassuming prairie vole, a small rodent—act remarkably like people. They form monogamous bonds. They "fall in love." They grieve over the loss of a mate. They become homesick. They have sex in response to chemical cues. They cheat on their "spouses." Males act like males and females act like females because their brains develop in deterministic ways, just as human brains do. And, as it turns out, some of the very same genes involved in these behaviors have now been shown to influence human behavior, too.

Of course, we're also going to make use of the latest findings from human tests and experiments. As you'll see, it has recently become possible to manipulate human emotions with the same neurochemicals scientists study in animals.

As important as human romantic love is, the stakes involved in what you are about to read extend far beyond romance, into the core nature of our societies. What social neuroscience is telling us about love also applies to how we live the rest of our lives and to the kind of world in which we'll live them. Human disorders like autism, social anxiety, and schizophrenia all feature elemental deficits in social interaction. Such disorders erode the ability of one person to socially engage. Since societies and cultures are built with the bricks of social engagement—from the first gazes mothers share with their babies, to the handshakes and smiles between shoppers and clerks, to a couple's first kiss—anything that weakens those bricks could have the same powerful effects on a society that it has on individuals.

Saying we're going to write a grand, unified theory of the brain, sex, and love—that we're going to try to answer the questions both ancient philosophers and Cole Porter have whiffed—makes our knees shake a little, in part because some of what you are about to read may prove to be controversial. It's important for us to stress that part of what we've written is argument, a group of hypotheses for a model of love. The hypotheses are based on science but are not themselves settled scientific fact. Still, we think this book is a bold attempt to explain the previously unexplainable. Ultimately, critics and readers will decide whether or not we've met our goals. At the very least, we think you'll come away understanding a lot more about love—about why it's not really crazy at all but the way we are made to operate. We do admit, though, that such knowledge might offer little solace when you wake up one February morning to find yourself in Minnesota.

1

BUILDING A SEXUAL BRAIN

A little over sixty years ago, in her book *The Second Sex*, Simone de Beauvoir famously wrote that "one is not born a woman, one becomes one." In the years since, Beauvoir's epigram has been turned into an all-purpose bumper sticker, embraced by feminists and fashionistas alike. Of course the fashionistas may not fully grasp Beauvoir's point. She was arguing that women have gender-based behavior thrust upon them by a patriarchy, whereas fashionistas tend to think that womanhood is bestowed by a slinky Ralph Lauren gown and a pair of heels, but the general idea is the same: one's behavior as man or woman derives from some external power. As the children of the small Dominican Republic town of Las Salinas proved, however, Beauvoir was wrong.

Luis Guerrero had no intention of rebutting the great French intellectual; he was simply intrigued by a riddle. As a young doctor working in a Santo Domingo hospital in the late 1960s, he encountered the peculiar life histories of some Las Salinas children and wondered why girls were turning into boys.

A native of the Dominican Republic, without many advanced research tools at his disposal, Guerrero couldn't do much to investigate the phenomenon he had encountered. But he packed his curiosity with him when he came to the United States to take up a residency in endocrinology, including work at Cornell University Medical College (today's Weill Cornell Medical College in New York City). Guerrero, who is now a doctor in Miami, helped interest Cornell researchers in making a trip to Las Salinas to see for themselves what was happening.

The 150-mile drive to Las Salinas from the Dominican Republic's capital, Santo Domingo, was a white-knuckle affair. In the early 1970s, the roads were mostly dirt. "These eighteen-wheelers would come screaming around bends," Guerrero recalls. "It was awful." Las Salinas was poor. Roofs on the houses were made of palm fronds. The main street, Calle Duarte, was a dusty track. Indoor plumbing didn't exist, and some homes didn't have outhouses. People bathed in a nearby river. The men who did not work in the salt mine that gave the town its name either cut down trees to make charcoal for cooking fires or farmed small plots of land.

Even today, there isn't much to attract visitors. The nearest Caribbean beach of the sort that has made the Dominican Republic a favorite of international vacationers is about fifteen miles away. A cemetery forms the western edge of town. Beyond that, the old salt mine gapes open, a briny slash in the landscape. Calle Duarte is paved now, and most homes have tin roofs and basic plumbing, but not much else has changed.

The Cornell research team soon discovered that the two dozen affected girls they found had been born looking for all the world like females. They had female-appearing genitals, complete with labia and clitorises. Naturally, their families raised them as girls. As these children grew, they wore bows in their hair and dresses—if they had any. They did the domestic chores girls in Las Salinas were expected to do while the boys roughhoused together and had fun exploring away from home.

Then, at around the age of puberty, they grew penises. This had been going on for generations, so long that the locals even had a name for it: *guevedoces*, or "penis at twelve." They called the changelings *machihembra* ("first woman, then man"), and, indeed, the girls did appear to become men. Their labia turned into scrotums filled with testicles. Their voices deepened and they grew muscles—a picture of one *machihembra*, around nineteen years old, shows the carved physique of a strutting middleweight boxer.

These youths' behavior changed, too. They walked with a macho bearing, joined the village boys in male play, and eventually started chasing girls. Most married. Some fathered children. Transition into male adulthood was not always easy, and there were lifelong differences between the *machihembra* and other men. The *machihembra* had penises that were somewhat smaller than average, and they didn't grow much of a beard. With the passing of years, they didn't develop a receding hairline. They also endured some social stigmas: imagine the taunts an adolescent schoolboy might suffer if his fellows knew he'd once been a girl. Even so, after *guevedoces*, the *machihembra* were fully male. More importantly, they accepted themselves as male.

When the Cornell team eventually solved the mystery of the *machihembra*, its resulting research article was given marquee treatment, appearing in the December 1974 issue of *Science*, the premier American scientific journal published by the American Association for the Advancement of Science (AAAS).

Almost exactly one year before the *machihembra* study came out, a man named John Money stood in front of an audience at the annual AAAS meeting to report on an explosive experiment he'd been conducting. Back in 1955, Money had declared that, as far as gender was concerned, babies were born blank slates. They might come equipped with the 46 chromosomes—including two Xs for females and one X and one Y for males—of one gender or the other, and the genitals of a baby boy or a baby girl, but, in an echo of Beauvoir, Money argued that this biological sex did not dictate one's "gender identity," a term he coined. Like Beauvoir, he insisted that gender behavior was *imposed* by parents, society, and culture. It was a position of nurture over nature.

For a variety of reasons, roughly one in a thousand babies is born in the United States with ambiguous genitalia. A girl may be born with an enlarged penislike clitoris; a baby boy may have a micropenis or no penis at all and undescended testicles. Babies are rarely born as true hermaphrodites with the reproductive structures of both sexes. What to do about such cases had always been a puzzle; the notion of doing nothing seems not to have had many advocates. By 1973, however, Money's views had become widely, even eagerly, embraced. This was partly because his power base at the prestigious Johns Hopkins University lent his views great weight. But he also salved a lot of anxiety among parents and doctors.

Surgeons who operate on children with ambiguous genitalia have long had a saying: "It's easier to make a hole than a pole." Constructing a penis for a 46,XX baby with a greatly enlarged clitoris, or making one for a 46,XY baby with a micropenis, or for a baby whose genitalia are so ambiguous as to be unidentifiable is extremely difficult. Removing a micropenis and constructing a pseudovagina is simpler—and that's what many doctors did when they surgically assigned gender to babies. Money's argument that this was a fine idea, as long as you followed up with lifelong hormone treatments and rigorous social and parental reinforcement, provided the intellectual justification doctors and parents needed. Few people chose to doubt him.

Though Money was sure of himself, and though his theory had been put into practice, there had never been a definitive test to prove that society creates gender identity. How could anybody possibly set up such an experiment? Ideally, one would have to use a baby with a perfectly normal chromosomal arrangement and perfectly normal genitals, then switch their genitals to those of the opposite gender—all of which would be wildly unethical. Even then, there would be no control (i.e., an untreated baby living in the same environment) against which one could gauge success.

The Reimer family walked into this void. Bruce and Brian, Canadian identical twins, were born in 1965 as two completely normal boys. After a botched surgical correction of the infant Bruce's foreskin destroyed his penis, the boys' parents consulted with Money, who immediately realized that Bruce's misfortune presented the ideal controlled experiment. Bruce and Brian had the same genes, had developed in the same womb, and would grow up in the same household—and because Bruce had been a perfectly normal boy until the surgical accident, there was no question of his "maleness," as there might have been had he been born with ambiguous genitalia or as a hermaphrodite. If Bruce grew up to behave like a typical girl, while Brian went on to behave like a typical boy, nobody could question Money's argument that society, not biology, exerts the strongest hand in making us behave as men or women.

The Reimers followed Money's advice. They had Bruce castrated and supplemented with estrogen. They raised him as a girl, changing his name to Brenda and giving Money the bombshell presentation at that AAAS meeting of what later became known as the John/Joan case.

The experiment was successful, Money told his AAAS audience. The male brother, Brian, behaved just the way we have come to think most eight-year-old little boys behave, he said, describing him as made of snips, snails, and puppy-dog tails, and as enjoying rough play. Meanwhile, Brenda, all sugar and spice, reveled in her little dresses and dolls.

As *Time* magazine reported after the meeting, Money's presentation provided "strong support for a major contention of women's liberationists: that conventional patterns of masculine and feminine behavior can be altered. . . . Money . . . is convinced that almost all differences are culturally determined and therefore optional."

In her 1898 treatise "Woman and Economics," feminist pioneer Charlotte Perkins Gilman declared that "there is no female mind. The brain is not an organ of sex. As well speak of a female liver." Now *Time* suggested that both she and Beauvoir were proven right. Second-wave feminists embraced Money's ideas as scientific proof that, above the shoulders, there were no important inborn differences between men and women.

The results of the Las Salinas investigation would seem to have thrown such conclusions into doubt. In all, the Cornell team found twenty-four men in thirteen different families who had experienced *guevedoces*. All but one of those families could trace their lineage back seven generations to a woman named Altagracia Carrasco, an indication that the condition had genetic origins.

Chromosomally speaking, the *machihembra* were normal males. They were 46,XY. They also possessed testicles at birth, but the testicles were undescended, tucked into their abdomens. What looked like labia was actually the raw material for a scrotum. The clitoris wasn't really a clitoris, but a penis awaiting instructions to develop—instructions that didn't arrive as the boys were gestating in their mothers' wombs. In other words, the *machihembra* were born pseudohermaphrodites. They looked like girls, but were actually boys.

The cause was a mutation, a misprint in a gene that carries instructions for cells to make an enzyme (a protein that catalyzes chemical reactions) called 5-alpha-reductase. This created a communication problem.

Cells are not self-starters; they require instructions. They receive those instructions through a system of signals—the signals are chemicals, such as hormones, that lock into receptors, or docking bays, located in or on cells. When activated by a docking chemical, the receptors for steroid hormones like testosterone and estrogen bind themselves to genes. This binding, in turn, activates (or inactivates, as the case may be) the targeted gene. If activated, the gene's recipe for making a protein is transcribed, like a ticker tape, onto a section of RNA. The recipe on the RNA ticker tape winds up at the ribosomes, the cell's protein factories. The proteins made in the ribosomes then head out into the world to perform a function. Because 5-alpha-reductase is a protein whose function is to turn testosterone into an even more powerful androgen called dihydrotestosterone (DHT), and because DHT signals cells to begin forming the prostate, penis, and scrotum, the bad recipe for 5-alpha-reductase meant that the fetal cells of the *machihembra* never got the message to start building male genitals.

Testosterone itself will dock on the same receptors as DHT—but not as efficiently. That's why, despite making testosterone as fetuses, the *machihembra* couldn't make enough to overcome the lack of DHT. But once they hit puberty, and got the massive rush of testosterone made in those undescended testicles (with a little help from their adrenal glands), the sheer amount of that hormone was enough to flood the receptors in the cells that are supposed to develop into the penis and scrotum. Voilà!— girls appeared to turn into boys.

DHT doesn't seem all that important after puberty, but some tissues, like the structures that make body hair and the prostate, still respond to it. In the *machihembra*, those cells only heard muted signals, which explains why the pseudohermaphrodites of Las Salinas had very little beard growth, small prostates, and hairlines that did not recede, even late in life. Hair follicles on a man's head can be sensitive to DHT. Depending on one's genetics, DHT sensitivity can lead to pattern baldness.

(Today, when you see a TV commercial with a middle-aged man doing the gotta-pee dance, or one in which an attractive woman runs her hands through some guy's verdant head of hair, you can thank the *machihembra* of Las Salinas. Drugs like Avodart, for enlarged prostates, and Propecia, for growing hair, are 5-alpha-reductase inhibitors.)

The Cornell team solved one mystery but only hinted at a solution to another. If Money was correct that society imposes gender identity and behavior, why, after appearing female for the first years of their lives, growing up as girls, and having a female gender reinforced every day for years, did the young men embrace their new maleness? Though they faced some difficulties, they didn't seem particularly shocked by their transformation. Something besides their new penises was telling them they'd been male all along. In fact, out of the original group of *machihembra* studied by the Cornell team, only one retained a female gender identity after puberty and, according to Guerrero, that exceptional individual was suspected of choosing a female identity as a ruse to gain access to young women for sex.

One year *after* the Las Salinas article appeared in *Science*, Money outlined a bright future for Brenda Reimer. "Now nine years old, she has differentiated a female gender identity in marked contrast to the male gender identity of her brother. Some of the other patients [whose treatment Money directed] are now adolescent or adult in age. They demonstrate that the twin can expect to be feminine in erotic expression and sexual life. Maintained on estrogen therapy, she will have [a] normal feminine physique and a sexually attractive appearance. She will be able to establish motherhood by adoption."

In 1979, famed sexologists Robert Kolodny, William Masters, and Virginia Johnson published their landmark *Textbook of Sexual Medicine*, in which they highlighted the importance of Brenda's transformation. "The childhood development of this (genetically male) girl has been remarkably feminine and is very different from the behavior exhibited by her identical twin brother. The normality of her development can be viewed as a substantial indication of the plasticity of human gender identity and the relative importance of social learning and conditioning in this process." Money's views had become medical gospel.

But that same year, one of the Cornell team members, Julianne Imperato-McGinley, wrote an article for the *New England Journal of Medicine* that followed up and expanded upon the original 5-alpha-reductase report. This time, there was no hinting. Imperato-McGinley argued plainly that "the extent of androgen (i.e., testosterone) exposure of the brain in utero, during the early postnatal period, and at puberty has more effect in determining male-gender identity than does sex of rearing."

Ruth Bleier, a physician and professor of both medicine and women's studies at the University of Wisconsin, as well as a renowned feminist crusader who founded a bookstore and café in Madison called Lysistrata (named for the heroine of the eponymous Aristophanes play who convinces Greek women to withhold sex from men), launched a scathing letter at the journal. Bleier had trained as a neuroanatomist at Johns Hopkins. Citing Money's research, she attacked the Cornell team's "scientific objectivity and methodology."

"The authors' failure to even consider an alternative explanation" for the switch of the *machihembra* from a female identity to a male identity was "truly astonishing," her letter said. Of course the girls had to act like boys, Bleier insisted: they grew penises! Everybody around them would start treating them like boys. To behave as a girl would defy all social expectations. Besides, she argued, girls in that society were restricted. They couldn't run around and play like boys; they were too busy doing domestic chores. Any sane person would make the calculation that it might be more fun to be a boy than a girl. "My fear," she continued, "is that this study, like others that incorporate preconceptions, faulty logic, and restrictive interpretations will be seized on . . . as demonstrating that our fetal brains are irreversibly imprinted, after all, by the presence or absence of androgens."

A few months after Bleier's letter was published, Brenda Reimer, then fourteen and desperate to live as a male, changed "her" name to David.

Money's grand experiment hadn't merely not worked—it was a disaster. Even while living as Brenda, young Bruce Reimer hated dresses. When Brian refused to share his toy cars and trucks, Bruce/Brenda saved up enough allowance money to buy his own. He bought toy guns so he could play army with Brian.

The truth was inconvenient, and not only for Money. Back in 1970, journalist Tom Wolfe had made fun of what he called "radical chic": left-wing politics promoted by the rich and socially connected. In the ten years since, radical chic culture had become mainstream, and one of its most cherished tenets was that recognizing inherent difference was prejudice. People were enamored with the idea of the "makeable society," recalls Dick Swaab, a pioneer in the study of brain and gender at the Netherlands Institute for Neuroscience. "Everything was 'makeable' and [Money's theory] fit into this concept." Brenda, now David, Reimer was a stark rebuke.

Perhaps that's why it took seventeen more years to debunk Money's story. It was not until 1997 that University of Hawaii sex researcher Milton Diamond and Canadian psychiatrist Keith Sigmundson (who had treated Bruce/Brenda under Money's supervision) published an article in the *Archives of Pediatrics and Adolescent Medicine* that shattered Money's triumph. Not only did Bruce/Brenda change his name to David, he eventually had surgeries to remove the breasts that had formed thanks to the estrogen he'd been given, and to create the simulacrum of a penis and testicles. He started taking testosterone, took a job in a slaughterhouse, married a woman, and helped raise her children. Sadly, he could not fully excise his tormented history. David Reimer used a shotgun to commit suicide in 2004. It was his third attempt. Even today, Diamond tells us, Money has adherents in the United States as well as around the world. His views are still reflected in some university gender studies programs that use clichés like "the social construction of gender."

/ Both the *machihembra* and David Reimer were always male because their brains were male, no matter what their genitals had to say, and all the socialization in the world was never going to change that.

THE ORGANIZATIONAL HYPOTHESIS

Cows don't often have twins. But if they do, and if the twins are both female or both male, the farmer gets a nice bonus. On the other hand, for hundreds—probably thousands—of years, cattlemen and dairy farmers have been disappointed by fraternal twins of opposite genders because the female is usually born a *freemartin*. Just how that term originated has been lost in a lexicological haze, but it was used as early as the 1600s to refer to a female calf born with a twin brother. She is almost always sterile; her brother is usually perfectly normal.

In 1916 and 1917, Frank Lillie, of the University of Chicago, began rooting around inside freemartins. He discovered that they often had ovotestes, an amalgamation of male and female gonads. They were hermaphrodites—the outcome of two different eggs being fertilized by two different sperm, and the resulting opposite-gender embryos fusing and sharing a blood supply. Male fetuses, Lillie realized, began making male-specific hormones (testosterone wasn't characterized until 1935, so Lillie didn't yet have a name for male hormones) before the female fetus's hormonal engine began pumping. Since they shared the same blood, she was getting a dose of those male hormones. She became masculinized.

Work like Lillie's spread the idea that hormones served important developmental functions in utero, but it wasn't until 1959 (five years after Money promulgated his theories on gender) that science began to understand how prenatal hormones could affect behavior. A research paper entitled "The Organizing Action of Prenatally Administered Testosterone Propionate on the Tissues Mediating Mating Behavior in the Female Guinea Pig" has the sleep-inducing understatement common to breakthrough science articles, but it became the foundation of what's known as the organizational hypothesis. Larry and others believe the events the hypothesis describes establish the brain circuitry that greatly influences all our basic love-related behaviors.

The outline of the experiment described in the paper was pretty simple. Charles H. Phoenix and his colleagues at the University of Kansas injected pregnant guinea pigs with testosterone, then waited to see what would happen to the babies. When the pups were born, the daughters of guinea pigs given higher doses of testosterone emerged with what a human doctor would call ambiguous genitalia. Phoenix later put those daughters into heat with hormone injections, then mimicked the way male rodents ask for sex, which consists of stroking a female's hindquarters; Phoenix called his imitation "fingering."

If she's in the mood, a female rodent will behave a lot like a swimsuit model during a photo shoot: she will arch her back and present her rear end, an invitation that signals she's receptive. This is called *lordosis*. When a soliciting male sees lordosis, he will mount. When Phoenix fingered the female guinea pigs whose mothers had received testosterone, they barely displayed any lordosis behavior at all. They did do a lot of mounting, though—about as often as males. The hormone hadn't just changed the daughters' bodies, it had changed their behavior, too. This means it changed their brains.

Phoenix tried giving testosterone to normal adult females, but it didn't have the same effect. Whatever was going on was happening in utero, during fetal development, to organize the brain in gender-specific ways. It also mattered *when* Phoenix gave testosterone to the pregnant mothers. He discovered there was a window: if he gave a dose of hormone when that window was open, he could lay down neurocircuits that would create male-typical behavior later in life, when neurochemicals activated these circuits.

The organizational hypothesis has held up remarkably well through years of refinements and additions. It holds that the default setting is on "female" during fetal development. At about the eighth week of human gestation, testicular cells will start making testosterone from cholesterol.

In a typical 46,XY fetus, testes begin to form and they make more testosterone. Later, they will get a boost from the adrenal glands that also make a small amount of testosterone. This testosterone and other androgens that derive from it, like DHT, spark the construction of male genitals from the raw materials available.

Testosterone, typically through conversion to other hormones like DHT (and even estrogen), also act in the brain to lay down male-typical neural circuits and permanently change the brain's chemistry. Later in life, when androgens are secreted, they activate these male circuits giving rise to masculine behavior.

In 1978, a UCLA postdoctoral student named Larry Christensen wanted to learn how to use an electron microscope better, so he reserved time on one for a practice session. He needed something to look at, of course, and because he was then working under Roger Gorski, who had spent most of his career studying hormones, gender, and the rat hypothalamus, there was a lot of rat brain around, especially slices of hypothalamus, a region at the very base of the brain that's a central player in regulating sexual, parental, feeding, and aggression behaviors, among others. It's also the brain's switchboard for hormone secretion from the pituitary and the gonads. Christensen mounted tissue on some slides, switched on the scope, and began looking. He soon noticed something surprising in the medial preoptic area, or MPOA: a spot on the very front of the hypothalamus above where the optic nerves cross each other (left eye to right brain, right eye to left brain) to form an X. (See Figure 1, page 19.)

Christensen found that one part of the MPOA seemed to be bigger in males than in females. He rushed to tell Gorski, but Gorski didn't believe him.

"He told me there were sex differences, and I said, 'Come on!' " Gorski recalls. "I was sure there were not any sex differences. I had looked at hundreds, if not thousands, of brains and never saw one."

A skeptical Gorski told Christensen to bring him the evidence. "Well, we had these two overhead projectors, and he projected them on the wall, and the difference just jumped out at you," Gorski tells us.

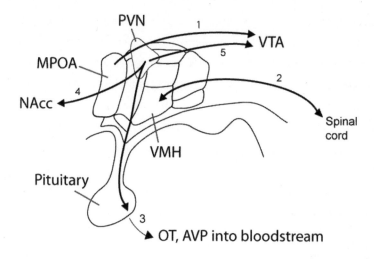

Figure 1.
The hypothalamus (left is forward). (1) The MPOA stimulates the VTA to release dopamine into the reward centers. (2) Sensory information from a courting male modulates VMH activity, which stimulates lordosis. (3) Oxytocin (OT) and vasopressin (AVP) from the paraventricular (PVN) are released from the pituitary and (4) into reward systems including the nucleus accumbens (NAcc), facilitating a bond. (5) OT neurons in the PVN also modulate the release of dopamine via connections to the VTA.

It was so obvious, that once Gorski and his crew knew what to look for, they didn't even have to magnify images of the region, which they later dubbed the sexually dimorphic nucleus (SDN). During all the decades of using rodents in labs all over the world, nobody had noticed this difference before, but as it turned out, the male SDN was roughly five times bigger than the female SDN.

In a series of experiments, Gorski proved that this SDN size difference depended on male or female hormone exposures in utero, just as predicted by the organizational hypothesis. He could give a single dose of testosterone to a pregnant rat and get a female pup with a male-sized SDN. In 1985, Dick Swaab announced that he had found the SDN in the human hypothalamus. It was two and a half times larger, with about twice the number of neurons, in men than in women.

The same kinds of manipulations Gorski performed in rats work in primates, too. One of Phoenix's students, Robert Goy, conducted experiments with rhesus monkeys, similar to the ones done with guinea pigs, and got similar results. Giving testosterone to pregnant monkeys created the same kinds of physiological and behavioral changes in the offspring. Even if a young monkey had female chromosomes, it acted like a male.

Goy did consider the issue of nurture. Did these females behave like males, he asked, because the other members of the troop, seeing a youngster with what looked like a penis, made the natural monkey assumption that it was a male and started treating it as such—thus suggesting that she'd better act like a male? If so, this would be a strong argument in favor of sexual "scripts" created by cultures to promote conformity to gender classifications.

To answer the question, Goy tried giving testosterone to pregnant monkeys at two different times. One group got the hormone early in the pregnancy, the other group got it late. The mothers who received doses early in pregnancy gave birth to females like the ones in the first experiment, with masculine genitals. But, despite having masculinized body parts, they played like their normal sisters and took on female roles. The babies of mothers given testosterone late in pregnancy looked just like typical female monkeys with all the proper female body parts. Clearly, the late testosterone boost missed the window to influence physical development.

Amazingly, though, these females played like young males—with more roughhousing and aggressiveness—but because they looked like females, their malelike behavior could not have been caused by monkey social pressure. There were clearly two windows of development that could be affected by androgens. One opened early in pregnancy and controlled the formation of genitals. The other opened later and organized the brain according to gender, masculinizing and defeminizing it in males. Without the testosterone surge, the brain would remain feminized—its default setting.

Larry's colleague Kim Wallen, who studied under Goy and is now at Emory University, has carried on this work, executing some of the most revealing experiments of all. He and his students and collaborators have used monkeys living in naturalistic settings—large social groups, intact families—so they're exposed to all the normal socialization monkeys experience. In 2008, Wallen and Janice Hassett used deceptively simple tools to test nature and nurture: toys.

Test after test has shown that if you put a typical little boy or a typical little girl in a room with a mix of toys—toys we usually think of as being for boys and toys we usually think of as being for girls—the boys play with the "boy" toys, like cars and trucks, and the girls play with the "girl" toys, like dolls. These results may not seem exactly shocking, but some have long argued that they prove that culture programs boys to prefer, say, a bulldozer, and girls to prefer a Barbie. After all, goes one argument, there were no Caterpillar tractors (and so no Caterpillar tractor toys) a million years ago, when we were still getting our evolutionary legs, so a preference for miniature versions of heavy landscaping equipment could not possibly be bred into our genes. Rather, under the influence of marketing, advertising, and social expectations, we cram our kids into gender categories when what we ought to be doing is setting them free from preconceived gender roles. Some parents choose to raise their children with this idea in mind, buying only "gender-neutral" toys or "girl" toys for boys and "boy" toys for girls.

Culturally speaking, monkeys don't have any particular toy prejudice—no Saturday-morning cartoons foisting *Star Wars: The Clone Wars* light sabers on males, no TV ads for females featuring Hasbro's Baby Alive Whoopsie Doo doll, who "'eats' and 'poops' just like a real baby!" The other young males in the troop aren't going to call Jake the rhesus a sissy if he picks up a doll and plays with it, and monkey parents don't particularly care what toy a youngster uses. Yet, when Hassett and Wallen presented monkeys with toys—specifically, plush dolls that included Winnie-the-Pooh, Raggedy Ann, a koala bear hand puppet, an armadillo, a teddy bear, Scooby-Doo, and a turtle, as well as a selection of wheeled toys that included a wagon, a truck, a car, a construction vehicle, a shopping cart, and a dump truck—73 percent of males preferred trucks and cars and 9 percent preferred plush toys. High-ranking females showed a pronounced preference for the dolls, as did a third of lower-ranking females. (Females were somewhat more flexible than males, with about a third preferring the wheeled toys. Another third didn't seem to care either way.)

The results clearly demonstrated that socialization does not dictate toy preferences. Those preferences are wired into our brains. Indeed, from the very first day of life, most baby girls prefer to look at human faces, whereas most boy babies prefer mechanical objects. In 2010, just in time for the Christmas doll-buying rush, Harvard University researchers announced that they'd been studying a group of wild chimpanzees in Uganda and had found the young females treating sticks the way human girls would treat dolls. The young females picked the sticks up and cradled them, taking them into nests and playing with them like a mother chimp plays with a baby. Young males never did this.

Monkeys aren't people, of course, and neither are guinea pigs. In the natural world, animals don't get shots of testosterone. But nature itself experiments with animals and with people, and those experiments have yielded powerful evidence to support the organizational hypothesis that gender behavior is built into our brains by the actions of hormones.

Africa's spotted hyena, for example, is that rare social animal in which females are dominant while males are submissive. Female hyenas are more aggressive than males, and the most powerful female hyenas are more aggressive than the other females. The females rule through intimidation. High-ranking females not only gain more access to more food, they exert breeding rights and have more offspring. In short, they behave like male lions or gorillas.

Female spotted hyenas are unusual in another way. They have penises— or what look like penises, but are really just incredibly large clitorises. These are so big that most untrained people can't tell a male from a female. They also have no functional vagina, giving birth through the penislike clitoris. This is a very bad way to give birth because it makes for such a long, painful delivery that babies can wind up suffocating before they reach the outside world. The reason for the female dominance and the malelike genitals is that hyenas are exposed to a lot of androgen during fetal development. Because the highest-ranking females receive the biggest doses of androgen as fetuses, and then go on to have the most babies, evolution is continually selecting for more and more masculine female hyenas.

Nature's human experiments can be equally as dramatic. About 1 in 20,400 baby boys is born with a condition called androgen insensitivity syndrome. Simply stated, they don't have any effective androgen receptors. Even though they make androgens, like DHT, there are no cellular docking bays to receive the instructions those androgens transmit. While they are 46,XY males, those with complete androgen insensitivity are born with female-appearing bodies and undescended testicles. This sounds like the *machihembra*. Unlike the affected children of Las Salinas, however, boys with complete androgen insensitivity exhibit feminine behavior when they're young, prefer girl toys and feminine play, and are attracted to males when they grow up. As far as their brains are concerned, they are women.

(Feminist writer Germaine Greer argues that 46,XY people with androgen insensitivity are men pretending to be women. She's wrong.)

One in 15,000 babies born around the world has a condition called congenital adrenal hyperplasia, or CAH. (The rates vary significantly from country to country. The rate in Japan is 1 in 21,000, while among the Yupik of Alaska it's 1 in 300.) CAH comes in several forms and degrees, but it's usually caused by an enzyme deficiency that leaves the androgen producers on the adrenal glands stuck in the "on" position. This gives a developing fetus loads of androgen hormones like testosterone and androstenedione (the 1990s hormone of choice of baseball players like Mark McGwire). Boys with CAH are physically affected—they are prone to short stature and infertility, for example—but their behavior is that of typical boys.

The effect of CAH on a 46,XX baby girl is more profound. CAH can result in ambiguous genitalia, an enlarged clitoris, and the appearance of a scrotum. She may have acne, excess body hair, and male-pattern baldness. A female might be infertile. Girls with more complete forms of the condition self-describe as lesbians later in life at a higher percentage than women who aren't affected, but even CAH girls who are less affected and consider themselves feminine and heterosexual play more like typical boys than girls.

As in the case of the Las Salinas children, these kinds of hormonally dependent, in utero developmental situations were once viewed with suspicion and carried a stigma. A talk given in 1918 by Dr. C. S. Brooks, of Frederick, Maryland, to the National Medical Association provides stark testimony as to the kinds of prejudice affected people faced. At the very moment he was speaking, Frank Lillie was solving the riddle of the freemartins, but too late for Dr. Brooks to understand its implications. He called his talk "Some Perversions of the Sexual Instinct."

A degenerate is an individual that is defective, either physically or mentally. I have observed three cases that were physically defective. In one case, a male subject, the genitals were very small and the urinary meatus opened on the perineum about half an inch above the anal orifice. In another case the urinary meatus opened just above the symphysis pubis. In a third case, a female subject, there was an infantile vagina, and the clitoris was elongated to the length of the Index finger. These three cases were sexual perverts, and their various anomalies could have been remedied by surgical interference, thereby restoring them to a normal place in human society.

These people were not perverts, of course. In all these cases—and in the case of all of us—a single molecule, a hormone, applied or withheld during discrete moments in fetal development, has not only affected genitals but also set a path for some of our most important behaviors over the course of our lives.

DIFFERENT CIRCUITRY, DIFFERENT BEHAVIOR

While the fundamentals of the organizational hypothesis explain the basics of heterosexual gender identity and behavior, homosexuality and transgenderism would seem to present problems. Most homosexuals, for example, do not have CAH or any other condition. Yet they behave differently in one of the most basic realms of human existence: they prefer to have sex with people of the same gender. Transgendered people usually don't have obvious medical conditions like androgen insensitivity, but they have a strong desire to change their gender.

Many regard homosexuality and transgendered people as notorious examples of how our culture has perversely allowed—even encouraged—people to reject the moral and social boundaries established by religion or custom in favor of aberrant ways of living, but gender-bending is more common in nature than you might think.

Some fish—groupers, porgies, bluehead wrasses—live as transsexuals. They change, typically from female to male. All bluehead wrasses, for example, are born female. They stay that way until a resident male disappears or dies. Then, the most dominant females immediately begin acting like males. They battle for preeminence. When one finally takes over, her ovaries degenerate, testes develop, and her brain creates new behaviors. She becomes a he.

Other fish are cross-dressing transvestites. Male round gobies, a species that has invaded the Great Lakes, guard their nests. But when they're on guard duty, they can't feed as well as they do when they roam, and finding a mate is a challenge because they basically have to wait around and hope a female shows up at their place. Some males, called sneakers, have devised a trick to take advantage of these limitations. Whereas the nesting homebody males look typical for their species—large, black, wide heads—the sneakers are smaller, mottled brown, and have narrower heads, like females. What's more, they behave like females. While the alpha male protects his bachelor pad and the Ferrari parked in the garage, along with any female who comes by to lay eggs, the meek cross-dresser slips in to propose a sexy ménage à trois. As males are wont to do, the alpha male finds this prospect irresistible. But the sneakers don't just have testes. They have bigger gonads than their brothers, and they make more sperm. So while the resident male thinks he's getting a three-way, the sneaker is actually fertilizing eggs left by a real female.

Many mammals, including primates, engage in homosexual behavior. The argument over same-sex coupling has always been whether or not any of these animals prefer gay to straight sex. What's true is that primates like gay sex. They have orgasms with same-sex partners. They ejaculate and make happy noises when they do. Dominant silverback gorillas have boyfriends. Some male langurs spend about 95 percent of their sex lives having homosexual encounters.

In part because monkeys and apes can't tell us, nobody knows for sure if any of these animals would pass up straight sex in favor of boy-boy or girl-girl sex. Many homosexual interactions in primates appear to be rooted in either necessity (males in all-male bachelor troops who have no other outlet other than masturbation, like guys in prison) or social communication (displays of dominance, conflict resolution). But because the behaviors exist and are so common, it's obvious that they are the natural products of the animals' brains.

A gay preference can be induced in individual animals. A phenomenon called the position effect, discovered by Fred vom Saal at the University of Missouri, works something like the process that creates freemartins. A female rodent that develops in the mother's womb between two male littermates can be masculinized and defeminized—turned into a butch mouse. Such females prefer to mount members of their own gender, display male-like behavior, and are less attractive to males.

Clearly, brains have an intrinsic capacity to display homosexual, bisexual, and transgender behaviors. Whether that capacity is expressed depends upon how a brain is shaped during development. Oregon Health & Science University biologist Charles Roselli believes that, when it comes to sexual preference, brains have a kind of "disinhibition" switch. The double-negative prefix makes for inelegant grammar, but it concisely describes what Roselli thinks happens in some animals.

A typical mammal's brain is organized to inhibit the desire to mate with a member of the same gender. Obviously, there's significant variation in the degree of this inhibition, or those male langurs wouldn't spend most of their lives having sex with each other—but the inhibition sets up an information flow within the brain telling an animal that sex with a member of the same gender is less desirable than sex with a member of the opposite one. Male-male sex may do in a pinch, but male-female sex: that's the good stuff. The switch can be manipulated, whether by a chemical, or the position effect, or human CAH. Sometimes individuals can be naturally and completely disinhibited, their capacity for same-sex attraction fixed in the "on" position. Roselli has come to this conclusion after spending the past fifteen years studying the only mammal other than humans known to prefer homosexual sex: male sheep.

For most of the year, ewes want little or nothing to do with rams. Then, in the autumn, as days shorten and the nights cool, the ewes come into heat. The aroma of their urine and their lady parts changes, becoming more appealing to rams than anything Coco Chanel could ever have created. Rams react as you might expect: they get very horny.

Luckily for the rams, the same hormonal signals that change a ewe's body also act as neurochemicals to change her behavior. Now she is receptive to the ram version of foreplay: bumping, sniffing her nether regions, kicking her legs, licking her. All the while, she will fan her tail to waft that sexy perfume his way, and often look over her shoulder at her would-be lover. Finally, she'll remain still while he mounts. Six months later, the farmer has a new lamb.

But some rams just won't cooperate. In agricultural circles these rams have been called "nonworkers" (farmers aren't romantics), or, as in a 1964 examination of the problem reflecting the Freudian lingo of the time, "sexually inhibited."

Roselli thought the problem of the uncooperative rams was an interesting one. He had come to Oregon to work with John Resko, a colleague of Phoenix and Goy's, but "I cannot say I came to study this particular issue," he recalls with a laugh. Nonetheless, Roselli began collaborating with scientists in Idaho who'd been studying rams they'd found to be "male-oriented." His job was to figure out what was going on in the brains of those rams.

One obvious idea was that gay ram brains somehow hadn't been defeminized like typical males. But these rams did not act like females; they acted like virile rams are supposed to act. They didn't stand still, waft their tails, and glance backward like ewes in heat. They sniffed, kicked, licked, and vocalized like any male seducing a female—they just did all those things only with other males. What's more, meeting a potential male sex partner created the same kind of hormonal surges in male-oriented rams as it did in female-oriented rams. It was just that ewes in heat, with their alluring scents and tail-wafting and sexy glances, elicited no chemistry at all in gay rams.

"You can't even give them female hormones and expect them to display female behaviors," Roselli tells us.

The reason why, he believes, is that the brains of these rams have been organized to be gay. He has come to believe that there aren't just one or two windows of fetal brain organization, but many that affect discrete brain regions at different moments.

The natural question, of course, is whether gay rams tell us anything about gay people. Roselli is cautious, saying, "It helps us with the questions. It's interesting," he adds, that "there are parallels. . . . Biology is biology: there is a common evolutionary thread there, and it gives me greater confidence in what science says about humans, but I think we need to do the science in humans, ultimately."

This is where Dick Swaab comes in. Enter his lab on any given day, and you might be reminded of a New York City deli: somebody is bound to be using the laboratory equivalent of a meat slicer to turn human brain into what looks like shaved ham. Slices of brain a few centimeters long and a micron thick lay on tray after tray for drying and storing, the trays stacked one atop the other. Before they're sliced, the brain bits—often from the hypothalamus—are kept in small jars, the jars filed away in metal drawers. A room full of long freezers holds more tissue. There are scores of drawers, hundreds of samples, and thousands upon thousands of slices, all because the Netherlands has made it easy to donate your brain to science.

Making a deposit at the Netherlands Brain Bank, which was founded in 1985 by Swaab, is as easy as signing your name, and then dying. If you die in the Netherlands, an on-call team (every hour of every day, no wait-ing!) will swoop in and remove your brain from your skull within hours of your demise. As of spring 2011, the bank, with about 2,500 brains, was one of the largest—if not the largest—resources of its kind anywhere. The col-lection took Swaab decades to assemble, but over the past twenty years it has helped scientists around the world issue a raft of findings.

Swaab himself makes heavy use of these brains. Today, he works in the modern concrete medical campus of the University of Amsterdam, about twenty minutes by tram from the central city. For twenty-seven years he directed the Netherlands Institute for Brain Research (now called the Netherlands Institute for Neuroscience), where the bulk of his fund-ing has been aimed at untangling the causes of psychiatric illnesses such as depression and schizophrenia.

In addition to the work that pays the bills, Swaab has what he calls his hobby. It's not really a hobby; Swaab takes this other work very seriously. But he has a dry and sardonic sense of humor, and when he says "hobby" he really means something like "very important science I've worked on for the past quarter century that hardly anybody is willing to pay for because it involves sex."

On the other hand, he has been assisted by those thousands of people who've given up their brains. You wouldn't think many people would rel-ish the idea of having their brains turned into deli meat, but since word got out that Swaab was especially interested in homosexuals and the transgendered, he's had lots of donors. One, an American who died at age eighty-four from lung cancer, struggled all his life with the feeling that he was a woman, not a man. He tried everything to conform to the life of a male. He married and had children. Throughout that life, he kept a me-ticulous private journal documenting his feelings and thoughts about his inner woman. Finally, he confessed his secret. With his permission, after his death his daughter had her father's brain removed and sent to Swaab, a postmortem effort to find out why he had lived as such a troubled soul.

As Swaab's funding difficulties suggest, the neuroscience of gender and sexual preference is not a popular line of scientific inquiry. But, in retrospect, Swaab was destined to pursue it. His father, Leo Swaab, trained as a gynecologist in Amsterdam. (He had barely finished his education when the Nazis occupied the Netherlands. As a young Jewish intellectual, Leo would surely have been targeted for deportation—relatives were murdered in Auschwitz—so he spent most of World War II in hiding.)

Like his father, and his grandfather, before him, Swaab went to medical school. He gravitated toward lectures given by an old family friend, a pioneering psychiatrist named Coen van Emde Boas.

When the Phoenix and Goy experiments were published, Boas latched on to them. In one of his lectures, the psychiatrist pointed out that the studies proved sex hormones were molding the brain during fetal development. Was it possible, Boas asked, that some hormonal action might be the cause of homosexuality?

Questions like that sent Swaab off on his so-called hobby. In 1989 he announced a controversial conclusion: The brains of homosexuals are different. They aren't different because the people in question are homosexual; such people are homosexual because their brains are different. The brains of the gay men Swaab examined weren't female. And they weren't exactly like heterosexual males, either.

The differences are caused, Swaab believes, not by a lack of hormone but rather by the timing of prenatal exposure. "It might be the moment the sex hormones are influencing the brain," he explains. This timing, he suggests, leads to differences that are neither male nor female in the way we usually mean those terms "but different. It's different."

Swaab has also found the transgender brain. After years of searching, his lab discovered differences in a structure called the bed nucleus of the stria terminalis (BNST). In lab animals, the BNST shares circuits with the amygdala and the MPOA of the hypothalamus. The BNST is also loaded with hormone receptors. It is sexually dimorphic like the rat, sheep, and human SDNs, and it is a key player in masculine sexual behavior.

When Swaab's lab looked at the BNSTs of male and female heterosexuals and homosexuals, and at those of male-to-female transsexuals, it found that the transgender BNST was the same size as the BNST of women. This was true whether or not the transgendered people had their testicles removed. These findings were used in the European Court of Human Rights to allow transsexuals to change their birth certificates and passports to reflect their brains' gender.

Other researchers have found a number of structures and circuits that differ in transgendered people. For example, in 2008, Swaab's lab reported that a structure called the interstitial nucleus of the anterior hypothalamus (INAH), which some believe to be the same structure as the human SDN, was also female-sized in male-to-female transsexuals.

Roselli has been trying to compare the INAH of sheep to that of humans. Again, he's wary of making any direct comparisons. Instead, he says "the best we can do is wave our hands and say it is in the same general area of the brain. It is my working hypothesis" that the sheep INAH and the human INAH are essentially the same thing and share some of the same functions.

For the most part, the transgendered people Swaab has studied haven't been gay. In other words, if they were male-to-female transsexuals, they were attracted to men. The brains of male-to-female transsexuals are not entirely female, however. In a functional study, when transsexuals were exposed to what was believed to be male and female pheromones, the brains of the transsexuals appeared to occupy an intermediate realm between straight men and straight women.

Nobody knows what precise mechanism creates transgendered identities, or if there is just one associated cause or a group of causes. But, whatever the cause, the effects appear as a range, and as subsequent research in Swaab's lab, and elsewhere, has shown, it's not just one structure that shows differences—there are differences across entire networks.

The same organizational mechanisms act in typical heterosexuals. In 2010, Simon Baron-Cohen, a professor of developmental psychopathology at the University of Cambridge (and cousin to Sacha "Borat" Baron Cohen), found that boys who had been exposed to more testosterone in the womb played more aggressively than boys who'd been exposed to less. These boys did not have a medical condition like CAH, and they weren't gay. They were just typical boys well within a normal range of testosterone exposure and behavior. But the differences still mattered.

Swaab admits that there is still much to be learned about how the brain creates gender. But he has little doubt that "gender identity and sexual orientation are both determined by the interaction between sex hormones and the developing brain." As Goy showed, the hormonal actions that trigger the formation of our genitals occur early in a woman's pregnancy. The hormonal influences that shape our brains happen later in a pregnancy. The two events can become disassociated, thus creating transgendered people.

Whatever the details, Swaab is insistent that the penis does not make the man, that the vagina does not make the woman. "If you look at the sex organs, you cannot make a conclusion about the direction the brain has taken."

As you can imagine, not everybody welcomes ideas like Swaab's and Roselli's. But if we accept that we are our brains, then a 46,XY human being with complete androgen insensitivity is a woman. If a person with a penis, a beard, and big muscles is unshakable in his belief that he is supposed to be a female, wants to act as a female, and is attracted to men, that person is a heterosexual female, not a gay man. A man who's attracted to other men but behaves and feels like a man is indeed a man; he just happens to have the brain of a homosexual. What the organizational hypothesis and work like Swaab's and Roselli's teach us is that genitals, and even chromosomes, can be irrelevant to gender behavior. Heterosexual boys don't try to turn a can of spray cologne into a flamethrower because they come equipped with penises or because their father plays catch with them in the backyard. They act like boys because they have boy brains. Heterosexual girls are typically more flexible in their gender expressions than heterosexual boys (Swaab believes women's brains are naturally somewhat bisexual), but a typical heterosexual little girl is far more likely to enjoy dolls and hold pretend tea parties, play dress-up, and, as we're about to see, fall for the "wrong" man because her brain is made that way.

2

THE CHEMISTRY OF DESIRE

Both our brains were organized similarly to those of many heterosexual males. So you probably won't be surprised to learn that Larry,
as he grew from being a little boy, began to eagerly await the arrival of the
latest Sears catalog because it contained actual, honest-to-God, not-
my-mom-or-my-sister photos of women in bras. When he was six or seven
or eight, Larry couldn't have cared less about photos of women in bras. By
age eleven or twelve, a nice demi-C cup—or, rather, the woman in the nice
demi-C cup—was pretty fascinating to him. Likewise, there came a time
when Brian began to anticipate the arrival of *Sports Illustrated*'s swimsuit
issue in much the way Druids must have awaited a solstice. Neither of
these revelations ought to surprise anybody. In fact, they make us absolutely, boringly average: billions of boys with brains organized like ours
experience an activation of the sexual desire circuitry that was laid down
during development—a mad rush of hormones that portends the coming
of puberty. When it happens, Cheryl Tiegs in a bikini becomes incredibly
relevant.

For boys and men, this activation remains at a fairly constant level (as-
suming good health) until later in life when testosterone falls, but even then,
androgens help keep sex on the male brain. Girls and women experience
something similar leading up to the hormonal storm of first ovulation.

Just like boys, girls will fantasize when their brains are electrified by pubertal hormones. Unlike boys, however, their hormones rise and fall dramatically as their bodies prepare for pregnancy. As a result, females fantasize about sex a little less often than males do—though not *that* much less often. At puberty, these chemicals not only change the way we look at the world (or, at least, the sexually charged part of it), they also greatly influence the way we behave in relation to it.

Take the young woman we are calling Susan, for example. She's about to dump her boyfriend. She doesn't know she's about to dump him—she hasn't been thinking at all about dumping him—but in just a few minutes, that's what she is going to do, conversationally, at least.

She's twenty-one and pretty, about five foot five, slim but not skinny. Her blond hair reaches her shoulders. Susan has a friendly smile and an open face. She's a little busty, but she's no glamour girl or pinup queen. She's the kind of young woman young men ask out on dates in the belief that she might actually say yes. Her boyfriend did just that, and they've been a serious couple for a while now. But he has a defect: he's not here.

Susan has walked into a pleasant, comfortable-looking room in a building on the University of Minnesota campus. There's a desk in the room, with a video monitor sitting on it. A camera, attached to the wall just above the monitor, points at the desk chair where Susan has now been seated. A researcher thanks her for agreeing to help with the project, an examination of male communication styles in the context of dating situations.

Susan has already been through phase one of this experiment. About two weeks ago, she heard these instructions:

You will be interacting with two different study participants who have twins. You will see the other participant separately on this television screen and will have the opportunity to communicate with each of them via a special camera system installed here on the wall. The other person has an identical setup in the room across from the living room. After you communicate briefly with a member of the first twin pair, you will communicate with a member of a different twin pair.

Today, she's communicating with the brothers of the young men she met via video two weeks ago—the other halves of the twin-brother pairs.

The video system is unidirectional. When one of the young men is talking, he won't be able to see Susan. When she's answering his questions, she won't be able to see him—so she has been advised that it's best not to talk when he's talking.

These twins, she's told, are the subjects of the study: there's no pressure on her. All she has to do is sit and answer a few of the guys' questions. Later, she'll be asked to share some observations with the experimenters. There are no right or wrong answers. When Susan's through, she'll receive a small cash payment—about enough to buy a couple of margaritas.

The first young man begins speaking. He's white and has short hair: good-looking but not swoon inducing. He seems shy and a bit awkward. He is, however, very sincere—just the sort of nice fellow a girl wants to take home to Mom and Dad.

"I'm bad at talking about myself, and I'm even worse at talking to cameras," he says. "So, if this is all wrong, just tell me, and I'll start over, though I have no idea what else I'd say besides this, so I kind of hope it's OK.

"Basically, I'm just a normal guy—or, at least, I *think* I'm a normal guy. Or, I'm normal enough, I guess. I deliver pizzas for this really popular place in Dinkytown [near the university] at night and am trying to finish up my English major. I've never been good at the dating thing; I'm really bad at being 'cool' . . ." He does have goals: "I think I'm a nice guy, too, and I'm looking for a nice woman. I guess, together, we'll be two nice people, or something like that. I'm not looking for a fling, or anything like that: I'd just like to meet someone I have some connection with, someone who's serious about making a life together. I'd like to get married and have a family."

When he tries to segue into the prescribed questions researchers have scripted for him, he says, "I'm going to ask you some questions . . . I hope I don't screw this up, too."

The first questions are the sorts people who've just met, but are interested, ask each other in bars and cafés and offices all over the world: What does she like to do in her free time? What interesting things has she done lately? Gone anywhere fun? Then the questions quickly change focus, probing her own dating tactics.

"Imagine that you're out with friends, and you see a man you find very attractive. Show me what you'd do to get his attention." He asks about her dating competitiveness. "Imagine that you and two or three other women are all interested in dating the same guy. What would you say and do to persuade him to date you rather than the other women?"

Susan whizzes through his questions, giving one- or two-sentence answers, elaborating little. In answer to his question about where she's gone for fun, she answers that she went away last weekend with her "boyfriend." They had a nice time. She politely but clinically answers his queries about how she might flirt. And then it's over.

After a short break, Susan sits down again to meet the brother in the other twin set. He, too, is good-looking and white, with short, dark hair. But, unlike the Steady Eddie Susan has just spoken to, he is glib, funny, and bordering on obnoxious.

"Usually, I do studies where you have to figure out which button to push to get a food pellet, and, if you push the wrong button, you get an electric shock," he deadpans. "But this one is *really* good because I get to meet and talk with *you*." Rather than explain why she should want to date him, he says, "I'm going to tell you why you should *not* date me. Then, when I'm done, we can talk and really get to know each other better. You should *not* date me if you want a guy who will always be on time, or someone who always remembers every single special event, like two-month anniversaries. You shouldn't date me," he continues, going on to mention skiing down black-diamond runs and wee-hour pancake feasts, until he concludes, "and, most importantly, you absolutely should *not* date me if you *don't* want to be swept off your feet and have a romance so intense that you'll question everything you ever knew and possibly begin writing with your left hand."

This routine might seem cringe inducing to you, but when the young man asks the scripted questions, Susan is eager to answer. She takes the questions about her own flirting techniques as invitations to display them, flashing what psychologist Paul Ekman called the "Duchenne smile"—a sign of unself-conscious pleasure that involves the muscles around the eyes as well as those around the lips. Susan giggles and laughs at his banter, though she's not supposed to say anything when he speaks. She flips her hair, cocks her head, plays with her earring, and leans forward in her chair—a tactic, explains social psychology and consumer-marketing researcher Kristina Durante, who's running this experiment, "to draw attention to her chest," even when he's talking and she knows he can't see her.

Then, when he asks her about any trips she has taken recently, she says she went away with a "friend." And, so, her boyfriend is unceremoniously dumped into the conversational ditch.

Don't judge Susan too harshly. To paraphrase Jessica Rabbit, Susan isn't bad: she's just organized that way. She isn't even fully aware she's flirting, or that she just erased her boyfriend from her life. Nor is she aware just how dramatically her behavior has changed between men.

As you may have guessed—Susan—not the young men, is the subject of this experiment. There are no twins. The two young men are actors, each playing one set of twins. Two weeks ago when Susan sat down to speak to them, today's smooth operator, whom Durante has dubbed the "Cad," played the role of his putative twin, a slightly shy young man who, like today's Steady Eddie, was looking for a committed relationship. Today's Eddie played his own brother, another Cad, two weeks ago. Back then, Susan responded to both the Cad and the Eddie about the way she spoke with today's Eddie. Today, though, she was outgoing and flirty with the Cad and yet, aside from some slight hairstyling differences, today's Cad and last session's Eddie look the same because, of course, they are the same.

While the men haven't changed, Susan has. Today, she's just about to ovulate.

Other young women at or near ovulation show the same behaviors as Susan. They put their fingers to their mouths, they look from side to side, they adjust their clothing frequently when talking to the Cads. "With the Cad, it becomes almost a scene you would see in a bar," Durante tells us. Then she brings up the rodent version of female seduction: hopping and darting and lordosis. "It's not such a jump to say that this is lordosis—showing they are receptive to this man. It's not overt, but this is the human version of that."

Most women have a pretty good understanding of how hormonal tides can change their bodies over the course of a menstrual cycle. The pituitary gland releases follicle-stimulating hormone (FSH). FSH prompts follicles, the egg buds on the ovaries, to grow and secrete estrogen. Luteinizing hormone (LH) forces one follicle to release its egg—ready to meet sperm—and send it on its journey through a fallopian tube. Estrogen and progesterone surge during this time, causing the lining of the uterus to thicken in preparation to receive a fertilized egg. Ducts in the breasts dilate. Progesterone rises, then drops dramatically, but estrogen remains elevated. If the egg isn't fertilized, the thickened lining of the uterus will slough away during menstruation.

The time just before, during, and after ovulation is a brief part of this cycle. But it is the time—the only time—a woman can become pregnant, and her brain knows this. So these hormonal shifts don't just affect a woman's physiology; they work in her brain to influence behavior toward maximizing the egg's chances of being fertilized so it won't go to reproductive waste.

As we mentioned in the introduction, when Larry began his scientific career, he started with lizards. He found that if he gave female lizards estrogen, they would always give in to a male's courting and mate. If he gave a male lizard testosterone, the male would become so sexually motivated, the animal would try to have sex with Larry's finger. When he turned to a species called the desert-grasslands whiptail, he was even more impressed at the power of hormones.

All desert-grasslands whiptails are females. Normally this would be pretty bad news for a species, but they reproduce through parthenogenesis—they make clones of themselves. In more typical lizards, females don't start developing eggs until a male courts them, but, of course, these lizards can't wait for a male. In his doctoral thesis, Larry figured out how they get around this obstacle: they've developed bisexual brains. When a ratio of sex steroids is just right, she becomes randy and courts, even "mates" with, another female, stimulating that female to kick-start her eggs. When the ratio reverses in the courting lizard, she will, in turn, be pursued and courted. In other words, the hormones, acting on suitably organized brains, create the mating behavior of two genders in the same individual.

Such a mechanistic view isn't the one most of us prefer for something more often described as kismet, or magic. As far as we know, Mila Kunis has never uttered the term "neuronal projections" in a boy-meets-girl movie. But there's a strong case to be made that, when we use words like "kismet," we are reaching backward through time to construct a rationalization around actions we weren't fully aware of performing—actions that were heavily biased by molecular events in our brains' circuitry of desire. Patti Smith's dictum that "love is an angel disguised as lust" is correct, but lust itself can be disguised as, say, shopping.

Unless we're in Frederick's of Hollywood, we almost never think of shopping as an expression of lust. But in fact, sex is often a hidden motivator of how we spend money and of what we wear—not to mention its relation to the kinds of behaviors Susan is exhibiting.

We often deny this influence. For example, when women are asked if they feel any more sexual desire near ovulation, many say no. But when they're asked at around the time of ovulation to count the number of times they've had sex or initiated sex within the past few days, they list more episodes than they do at low fertility points in their cycles. At ovulation, they enjoy porn more than at other times of the month. They become favorably biased toward ruggedly handsome men, as opposed to pleasant-looking "nice guys." They tend to avoid their own fathers, consume fewer calories, and spend less money on food than on clothing and sexy shoes. Women also fantasize more often about having sex with someone who is not their current partner.

As Durante explains, Eddie may be loyal, hardworking, and looking for commitment, but those are qualities that appeal to one's rational brain—that part of ourselves that calculates the benefits of delaying reward for long-term benefits. Such reckoning takes place in the cortex, the biggest part of our brain, but hormones have harnessed other parts of Susan's brain and amplified their voices. Today, she's all about the short term, and an awkward, English-majoring pizza delivery guy who wants to get married is not satisfying her immediate need.

"What estrogen is doing to her brain is letting it know that the problem right now is mating," Durante says. "And all the energy is directed right there. This is not conscious thought." The brain knows that, for the just-released egg, it's now or never. And the Cad's glib confidence—his talk of skiing black-diamond runs, his implied promise of romantic thrills—makes him appear fit, vigorous, and dominant while Eddie exudes uncertainty. The Cad also has the advantage of merely being present and available: Susan's boyfriend is out of sight, out of mind.

The Cad may be a braggart, but he seems like a winner. And the hard truth for men is that no matter how nice a guy you are, at their most fertile moments females of all species appreciate winner types most. When Stanford scientists, led by Russ Fernald, looked at how female fish process social cues about the fitness of their mates, they discovered that when a gravid female (one heavy with eggs and ready to spawn; roughly the same as an ovulating woman) watched her preferred mate win a fight against another male, she showed increased activation of neurons in the MPOA. The ventromedial hypothalamus (VMH, Figure 1, page 19), a region that directly regulates female sexual behavior, was also activated. In short, when her favored mate won a fight, he instigated her reproductive and sexual attraction centers.

When her preferred mate lost a fight, however, her brain's anxiety-producing circuits clicked on, seeming to indicate she was experiencing stress. Put in human terms, she appeared to be worried that she'd picked a loser boyfriend to be the father of her babies.

Durante isn't kidding when she compares the behavior of the young women in her study to that of lab rodents. When Larry needs to use one of his female lab animals for an experiment or observation, and also needs her to be estrous, and to display the behaviors associated with estrus, all he has to do is give the animal an injection of estrogen, followed by progesterone, and he gets an almost instant change of behavior. As Durante says, the behaviors she sees from ovulating women aren't nearly as dramatic as lordosis, but they are parallel.

Susan's behavior is a more subtle example of what scientists call proceptive, or appetitive, behavior. A female rat's hopping and darting is proceptive. Susan's flirting is proceptive. Her hormonal ecosystem is creating an appetite for mating, even if she doesn't fully realize it and probably won't act on it—at least not with the Cad. When driven by these same hormones, some animals send out very obvious signals that they want sex, as many owners of female cats can report.

THE CATS WHO COULDN'T SAY NO

Let's say you have a cat named Buttons. While you're a conscientious pet owner, you haven't yet had Buttons spayed, perhaps because she's so sweet tempered, calm, and innocent that she's lulled you into complacency. Then, practically overnight, fluffy Buttons becomes a cat version of a drunken sorority girl on spring break, pawing at the carpet as if doing a backward moonwalk, spraying, rubbing her nether regions on the furniture, and yowling so loudly the neighbors complain. Buttons begins arching her back, pressing her chest to the floor, and pointing her hind end toward the ceiling, tail raised. The kids begin asking questions: "What's up with Buttons?"

What's up with Buttons is that she's become a very horny kitty and is displaying proceptive behavior to let all interested parties know she's in the mood.

Before, when Buttons was approached by a male cat, she hissed and tried to claw his face. But now she's in heat—and if she could get away with it, little Buttons would have sex with the first male, and the second and third and fourth, that managed to reach her. She'll be like this for about a week, or until she becomes pregnant. Then, just as suddenly as the bacchanalia began, it will stop.

This is a remarkable display. Naturally, scientists wanted to study its mechanisms. But studying lust proved to be much more difficult than anybody anticipated, partly because no one could figure out how to turn it off. They did their killjoy best. They cut the nerve signals to the erogenous zones of cats and rabbits and rodents, but the animals still displayed mating behaviors. They destroyed parts of the creatures' brains—like the olfactory bulb, so they couldn't smell a scent that might trigger sex behavior. They even lobotomized them and removed their eyes, but the animals continued to act out mating behaviors. In other words, researchers tried slicing away any enjoyment of sex, any stimulus from the opposite sex, any "reason" at all to have sex—and the lab animals still displayed an appetite for it. Biologists even wired cats up to see if they could juice lust: it turned out that shooting electricity into cat brains just gets you angry cats. Jolting the cats triggered "a variety of functions," Geoffrey Harris and colleague Richard Michael, of the Maudsley Hospital in London, dryly noted, "some of which (such as the rage response) may be antagonistic to the expression of sexual receptivity."

There seemed to be just one way to shut down sexual desire in female cats: removing their ovaries. Take those out, and a female cat will never mate—or show any interest in mating. After this discovery of the cat on-off switch, Michael removed ovaries from females, then administered a synthetic estrogen to replace what went missing. Regardless of time of year (female cats generally come into heat as days lengthen), he got lusty female cats. This may seem obvious, but think of what it means. Mating in mammals is a pretty complex behavior that involves two individuals interacting in very specific ways in response to each other. All kinds of variables can come into play—like the light-dark cycles of days, for example. Female cats have good reason to avoid intercourse: male cats have backward-facing, Velcro-like barbs on their penises. But Michael's results were so predictable he could mathematically chart the sole variable—estrogen dose—with the mating behaviors he saw. One molecule, delivered under the skin, initiated female-cat sex mania.

This raised the question of exactly how the estrogen was inducing proceptive behavior. Was it the physical changes of estrus, the preparations being made in the vagina and uterus for a coming pregnancy? Did those changes act like aphrodisiacs, creating an "itch" that wanted scratching? Or did estrogen act directly on the brain, apart from whatever effects it was having on the rest of the body?

Harris and Michael conducted a tandem experiment (supported by a grant from the US Air Force; one can only imagine why). This time, when they removed the cats' ovaries, they didn't replace estrogen with injections under the skin: they created implants that could put the synthetic estrogen right into the cats' brains—specifically, into the hypothalamus. With the implants, the female cats showed none of the usual physical symptoms of being in heat—no changes in the vagina, in the cells of the uterus, nothing. But they showed all the usual behaviors: the treading, the yowling, the rubbing. One group of implanted cats "appeared to be in a state of continuous sexual receptivity throughout the survival period. These females could be described as hypersexual, on the grounds that they would repeatedly accept successive males at any time of the day or night without showing any signs of the refractoriness which normally follows mating." They just kept having sex, over and over, for up to fifty-six days and nights.

The estrogen molecule was acting specifically as a signal in the hypothalamus to dramatically alter behavior without making any of the systemic physical changes that normally accompany estrus. Behavior didn't depend on changes in the reproductive tract. A hormone, acting as a long-range signal on brain circuits laid down during fetal life, instigated behavior.

Scientists like Harris and Michael can separate bodily changes from brain-driven behavior in the lab. In the natural world, a feedback system with the hypothalamus acting like a network telephone exchange "listens" for the physiological changes of estrus induced by ovarian hormones and then coordinates behavior so that a female can take action to become pregnant when the time is right. Donald Pfaff, a neuroscientist at Rockefeller University, spent nearly forty years working out precisely how the hormones, and a few other neurochemicals, activate specific circuits in rodents to create these behavioral changes.

Cats don't release eggs until they have actually mated, but most rodents are more like humans: they spontaneously ovulate following a surge in estrogen and progesterone in a four-day cycle, which is what Larry mimics when he gives injections to animals whose ovaries have been removed.

Estrogen exerts its powerful behavioral effects by docking at receptors concentrated in the ventromedial hypothalamus, just behind the MPOA. When these receptors bind to the regulatory regions of certain genes, the chemistry of the neurons changes. They increase neurotransmitter production and become more sensitive to other signaling molecules. For example, when estrogen docking has increased the production of progesterone receptors in the ventromedial hypothalamus, the resulting heightened sensitivity to progesterone allows the hypothalamus to time ovulation with amazing accuracy. This estrogen-progesterone sequence leads the female rat to start her seductive hopping and darting precisely at the time she is most fertile, and then to assume the lordosis position if a male takes the bait.

Just as they do in rats, such biochemical and neural changes prime women's brains to respond to social signals from potential mates in ways they don't when ovulation is absent. That's why Susan turned on her charm for the Cad today, but not two weeks ago. These changes are necessary because life is about more than just sex. (Really.) Rodent brains, like human ones, juggle competing demands for their attention. Changing the chemistry of the neurons helps divert a female rodent's attention away from foraging for food, worrying about snakes, or feeling pain—and toward sex. This shift is necessary because having sex is risky business for a rodent—it's difficult to run away from a rattlesnake if you are hopping and darting in front of a male. If she is worried about being eaten, she's unlikely to be interested in mating. Her sex drive is inhibited because she's got to attend to other business.

Recent studies have shown parallel systems at work in people. There is now strong evidence that, for women, proceptive behavior is intimately tied to the brain's stress response. The stress response—including those feelings of wariness, anxiety, edginess, perhaps in response to an untrustworthy man, or a bad day at work—is mediated by interactions among the hypothalamus, the pituitary, and the adrenals. This stress axis operates with input from the amygdala, the brain stem, and the prefrontal cortex. Regions within the hypothalamus communicate with the brain stem, the most primitive part of our brains. The brain stem activates the autonomic nervous system, which, in females, modulates ovarian functions and sexual drive. The amygdala communicates with hypothalamic areas in order to regulate hormones and reproduction; the prefrontal cortex directs focus toward a goal.

As you can tell from this list, there's a great deal of overlap between the circuits that make up the stress response and the ones involved in sexual behavior. Whether to start worrying, or to relax and enjoy some sexy time, depends largely on estrogen. If enough docks onto estrogen receptors (as happens during ovulation, when estrogen levels are high), a female's stress response is tamped down; her brain focuses less on things that could provoke anxiety, like the prospect of mating with someone who has Velcro on his penis. Men try to help this process along by lighting candles, opening champagne, and playing bossa nova. We might think we're being suave, showing off our moves, but we're really turning down the anxiety volume so a woman can hear her estrogen talking.

When the hypothalamus gets the message that all is well, and the ovaries are ready to release an egg, estrogen, functioning as a neural signal, shifts the bias of a woman's brain from avoidance to approach, to doing what it takes to get pregnant—the "problem" that Durante says Susan's brain is trying to address.

Several things happen when these desire doors open, whether one is a mouse or a woman. A mouse in heat, or an ovulating woman, can suddenly possess enhanced positive perceptions about the opposite gender. Any negative sensation like pain will be reduced. She'll be less vigilant for danger, more willing to take a risk. If the female is a rodent, she'll hop and dart in front of a male. If she's a woman, she'll be more likely to flirt.

Susan's brain is directing her attention away from the fact that the Cad is a bit of a jerk, Durante says. "There is an over-perception of Cads as great guys when you are ovulating, but only when it comes to you." Ovulating women often think, she says, that "of course, the Cads could not possibly be interested in other women!"

"This is really a balance of two brains," suggests Heather Rupp, a neuroscientist formerly affiliated with Emory and Indiana universities. "What changes across the menstrual cycle is the predisposition based on the ability to shut out the drive." Because Susan has a rational brain, she isn't likely to use her small payout from the experiment to invite the Cad out for margaritas and then take him to bed, breaking her boyfriend's heart. There would be costs to her reputation, to any future plans she's made with her boyfriend, to her own sense of loyalty. At or near ovulation, the bias may not shift far enough away from these costs for Susan to take a room at the nearest Motel 6, but it moves the needle significantly toward letting go. If the Cad happens to be the nearest desirable man, he gets the attention.

While at Indiana, Rupp placed women in a functional magnetic resonance imaging (fMRI) machine, a device that can detect brain activity and produce pictures of structures responding to stimuli. The women were imaged when their progesterone was high and their estrogen was low, and when the ratio reversed near ovulation. At both times, she showed them pictures of male faces to consider in the context of the following scenario:

You are not in a committed relationship and are open to a sexual encounter. You and some friends are out Friday night. While out, you meet the man presented in the image for the first time. You two have a good time talking together and that continues into the evening. You and he end up back at his place to continue hanging out. It is clear to you that he would have sex with you if you want to. Imagine that you are in this scenario and open to a sexual encounter. Based on the image and information presented, please indicate using the button box: How likely would you be to have sex with him?

Near ovulation, the medial orbitofrontal cortex—the area in the prefrontal cortex associated with the calculation of risk-and-reward decisions and goal-directed behavior—was significantly more active, possibly reflecting that the women were more seriously assessing the men as sex partners and, according to the hypothesis, more likely to have sex.

Importantly, the same region was also somewhat more active in response to another scenario: "Please make your decision regarding the likelihood of renting the house portrayed in the image based on how attractive you find it." Like having sex with a strange man, renting a house presents a risk. Ovulation lowers a woman's resistance to risk taking, just as an increase in testosterone lowers such resistance in men.

Rupp is reluctant to say that these mechanisms explain why some women become entangled with "bad boys." "*Bias* is a nicer word," she says, laughing. "People are less offended by that. But I do think women are driven and predisposed by their brains and hormones to act within certain windows of possible behavior, and this is why good women pick bad men. It's that mid-cycle preference and the way their brain prioritizes. The guy you are most likely to pick mid-cycle—he is not necessarily the guy who is going to raise your children. The perfect guy is the guy you like across the entire cycle, and they are rare!"

As Rupp and Durante both remind us, predisposition is not the same as predestination. For a female cat with an estrogen implant in her hypothalamus, sex is a foregone conclusion. If a female rat's seduction works, and a male touches her flanks, her hypothalamus, specifically her VMH, senses the massage and signals the brain regions that control motor function. Her muscles will almost instantly arch her back in lordosis in preparation for intercourse. (See Figure 1, page 19.) A woman's reaction is a little more complicated, but the same mechanisms influence her behavior, predisposing her to act on sexual impulses.

LAP DANCERS, MONEY, AND MEN

There isn't a good explanation for why men might pick the "wrong" woman. We mean to say there is an accurate explanation, just not a very appealing one: We're easy. We're generally quite happy to have sex, and we don't require a lot of clever repartee from a woman, welcome as that is, nor any advertising of her status or prowess in realms other than sex. We know this may shock and amaze women, but we're not all that choosy— especially when we're young, and our testosterone is at its highest.

The reasons are simple: sperm are cheap; we have millions of them, and our testosterone is a bit like the constantly circulating hot water in an apartment building. Our androgen receptors are the water taps in the hypothalamic "apartments" related to sexual behavior, and in the amygdala. There are many taps, they're always ready to be turned on, and since spending sperm costs us little, we'll splurge like Berlusconi on vacation.

Men evolved this way. Male mammals are usually unable to tell just when they might encounter a willing female, so they have to be ready—or easily made ready—by, for example, something as simple as hopping and darting.

Just how critical testosterone is to male sexual desire has been demonstrated time and again by many scientists. Using some of the same techniques as Harris and Michael, they've restored it in castrated animals simply by implanting testosterone in the MPOA. Capons started acting like roosters again. Mice and rats resumed marking their territory with urine, having erections, mounting, ejaculating. This was true even though the animals didn't show the usual physical effects of testosterone. Capons didn't grow bigger combs, and the rodents didn't grow bigger prostates. Like the estrogen in the female cats, the hormone was acting only in the brain, and only on very definite circuits.

For thousands of years, societies acted on the principle that lower testosterone reduced male sexual desire, castrating men to create eunuchs. The belief was that eunuchs, freed of carnal urges, made better and more trustworthy managers of households and imperial businesses and could be safely loosed among the women. But while castrating a man usually does render him impotent, it doesn't always—a little secret shared by some castrati and a few wandering wives.

Testosterone will drop in certain contexts besides missing testicles. When male marmosets sniff a test tube containing the scent of their own infants (but not strange infants), their testosterone drops within twenty minutes. This could be a protective mechanism, a way of tamping down a father's sex drive and aggression when he's around his babies. Studies in humans have shown that when men become fathers (and especially when their babies are newborns), they experience a significant drop in testosterone. Men who are very involved in parenting their babies show the biggest drops—probably because the act of caregiving itself lowers testosterone and allows the brain to focus more on nurturing and less on fighting (even if it's only a job promotion at stake), sex seeking, or anger. Testosterone falls when we lose a sports match, and it even slumps when the team we root for loses a game. (Sorry, Cleveland.) Driving an old, clunker family car reduces a man's testosterone levels. Long-married men have lower testosterone readings than single men do (read into that what you will). Being fired from a job isn't good for our levels, either. Though menopause reduces sex hormones more dramatically in women, aging men still experience a significant falloff.

On the other hand, context can also raise testosterone. Winning does it, whether the competition involves sports, a battle, a chess game, or a political campaign. This effect can self-perpetuate: a win imbues a man with confidence, further enhancing his chances of winning again. Driving a Ferrari will do it. When male mice catch the scent of an estrous female, their testosterone rises, their senses become enhanced, and they become extremely goal directed, following any female who does her hopping and darting dance of seduction and smells fertile. Something similar happens in monkeys. Testosterone spikes within thirty minutes after male marmosets are exposed to the scent of ovulatory females. The males also spend more time smelling the scent and have erections that last longer than when they sniff a control scent.

That last experiment is an example of just how interactive sexual motivation can be. Reproduction requires an exchange of information between would-be partners. A female rat hops and darts. The Cad brags about black-diamond skiing. Susan leans her chest into the camera. Just as Susan became more motivated when speaking to the Cad, men become more motivated in the presence of fertility.

When it comes to women, calling these behaviors "estrous" or saying that women go into "heat" is very controversial. For many years, most scientists thought women didn't display any obvious signs of their fertility. While everybody came to understand that hormones played a role in sexual desire, most scientists thought women had, at most, a "hidden" estrus. Unlike cats or rodents, humans will have sex any time of year and any time of day or night; we use sex for much more than reproduction. Somehow, went the thinking, we had become uncoupled from the basic contextual sex drivers operating in other animals.

A variety of evolutionary theories have tried to explain this. Some argued that women evolved to keep men guessing so they could more easily control when they had sex. Others claimed that women evolved to keep men guessing so they'd never know when women were fertile and, not wanting to take the chance of providing for another man's baby, would stick around to guard against other males. That's how we got monogamy, some say.

But what accounts for the behavioral signaling between two people that indicates they're willing and eager? How do they perceive that it's time? Why has Susan's behavior changed from two weeks ago? In fact, evidence is now mounting that women do have an estrus—and that men can sense it. This new evidence suggests that men like the subtle cues of human estrus so much, they'll pay to be near it.

The Spearmint Rhino is a temple dedicated to desire. But desire is merely the tool for advancing the temple's real business of separating men from their money. Located in a warehouse and light industrial district just beyond the Las Vegas strip, the "Rhino" is a disorienting maze of large rooms filled with men and female strippers. Though the Rhino is open around the clock, it's always midnight inside. During our visit, it's so dark that, at any distance greater than ten feet, the dancers become fuzzy outlines, illuminated only by dim pink and purple lighting. Bartenders in push-up bras navigate the booze by feel and memory as much as by sight.

As with any temple, the design aims to focus one's attention. Just as the nearby casinos strive to keep gamblers' attention on wagering, inside the Rhino the darkness, the maze of rooms, and the surreal safari-club-cum-bordello atmosphere create a sensory deprivation experience that obliterates any concerns a patron might have about the world outside—the convention he's supposed to be attending, the amount of money he's lost at the craps tables, the wife at home. Nothing exists inside the Rhino except the men, the dancers, the music, and the teasing sense that contrary to all experience you really *are* the kind of guy a hot woman in a schoolgirl micromini wants to take home.

If you want to indulge this fantasy, you must observe certain rituals. Dancers dressed in some variation of thong bikini bottoms, high heels, thigh-high stockings, and slinky club dresses dance on small platforms several feet higher than the surrounding semicircular arrangements of seats from where the men watch. Others cruise the rooms, chatting up customers. A dollar, or two, or five, placed on a table during or immediately after a dance is a signal that a customer might be interested in spending more time with the dancer. The same message is sent when a customer buys a drink for a stripper who has approached him.

"I'm a third-year medical student," a dancer, who says her name is Lana, tells us as she stands at the bar, sipping a drink we've purchased. Maybe Lana is a medical student, and maybe that's just her story. (If all the strippers who said they were medical students actually were medical students, there'd be a glut of doctors, and the attendees at annual AMA conventions would look like the cast of a Mexican *telenovela*.) Either way, Lana has obviously given her patter a lot of thought. She has also carefully selected her costume—garter belt, lacy bra, G-string, heels so high she towers over most of the men—for maximum appeal. She leans into Brian as she speaks into his ear.

These are tactics strippers use to entice patrons into paying for a twenty-dollar lap dance—or a series of lap dances costing a hundred dollars or more in a private "VIP" space. (Anyone who fails to "tip" for a lap dance will usually be reported to one of the linebacker-sized male employees who roam the club, and whose duties include strenuously suggesting adherence to protocol.) One lap dance could lead to another, and another, and possibly the purchase of extras, like a bottle of wildly overpriced vodka. The male customer may sense he's spending far too much money. But, by now, his amygdala and hypothalamus are being flooded with testosterone, and he's become very focused on the "problem" at hand.

The seduction may be ersatz, but the neurochemical cascade it provokes is real. Because men place value on obtaining wealth and displaying it when we are around attractive women (which explains the sales of Cristal), a male customer receiving a lap dance wants to act the big shot for his dancer. So he'll suspend disbelief and hand over money, sometimes lots of money: a tactic to advertise his desirability as a short-term mate, even though he knows the dancer won't really have sex with him.

Dancers can earn high wages this way, but first they have to get a man to accept the offer of a lap dance. Their ability to do so doesn't just depend on what they wear, or on the quality of their conversation. How much money a dancer makes can also depend on whether or not she's nearing ovulation.

"Men in these clubs are looking for the most hotness per dollar," Geoffrey Miller, a psychologist at the University of New Mexico who has tracked stripper income across ovulatory cycles, tells us. Given the inherent competition among dancers vying for male attention, where attention equals dollars, "the customer can use his own testosterone level, his own arousal, as a sort of indicator of women's hotness, and he is willing to pay extra for it, to be with the woman who pushes his buttons." Miller has found that the women who push male buttons most effectively on any given shift are women near or at ovulation.

The effect Miller discovered wasn't small or subtle; he calls it "shocking." Miller and his colleagues used New Mexico clubs as real-world labs to conduct a study they titled "Ovulatory Cycle Effects on Tip Earnings by Lap-Dancers: Economic Evidence for Human Estrus?" Based on the financial outcomes, it seems clear there's no need for the question mark. When the strippers Miller chronicled were in estrus, they made about $354 per five-hour shift. Anestrous women made about $264—a difference of $90. Menstruation cut dancers' earnings in half. The difference can't be attributed to one woman's attractiveness over another's, or to fashion choices, because the research took place over two months. The estrous and anestrous women are the same women, documented at different points in time. Dancers who were on the birth control pill—which essentially eliminates estrus—made about $193 per shift, also far less than ovulating women.

By linking dancer preference to money, rather than to customers' statements, Miller was able to show this preference was real and unconscious. Talk is cheap; if a guy opens his wallet, he means business.

Like the men, the dancers were completely unconscious of the effect. Some thought they made more money during menstruation than at any other time in their cycle. "They talk about their earnings a lot," Miller explains. "That's the major source of gossip; it's what they are there for—to feed the kids, to get tuition—but none of them made any connection. None of them realized being on the pill hurt earnings, and none of them scheduled their work shifts to maximize earnings."

But if the men preferred estrous dancers, how did they know which ones were in it and which ones were not? They didn't consciously pick estrous women over anestrous ones, of course. But the clubs Miller selected for study were dark and loud, designed to focus male attention on the dancers. By observing the women as they performed their stage dances, and then in close contact during the conversation period (the music can be so loud in these places you practically have to put your mouth on a person's ear to be heard), when they decided whether or not to pay for lap dances, men were able to pick up subtle cues. There was—literally—chemistry between the sexes.

When men have been exposed to odor samples taken from women at or near ovulation, they, like their monkey cousins, show a spike in testosterone, compared with men who sample odors taken from women who aren't ovulating. The close contact required for conversation in a noisy club might be giving the men olfactory hints.

Miller thinks other factors might be at play, too. "We know from other research that, with ovulation, the quality of the woman's voice—timbre and pitch—is more attractive. The quality of the skin, the facial attractiveness, increases a little in ways that are not well understood. Body shape changes a little—there's a relatively lower waist-to-hip ratio—and some studies have shown that women at peak fertility have higher verbal fluency and creativity."

Not only do men pick up on these cues, but Miller also suspects that estrous women behave differently. With the lowered sense of anxiety and risk that accompanies ovulation, estrous women may be more willing to make an approach. They may have more confidence in their own sex appeal and bodies, and may move more seductively when they perform.

Put another way, estrous dancers are better at displaying proceptive behavior—their version of hopping and darting—and the men, responding in kind with a rise in testosterone, become more focused and goal directed. This, in turn, drives them to open their wallets.

Miller says this is a plausible theory, adding that, although the dancers seem unaware of the power of estrus, they manage to take advantage of it anyway. The reason most men can't stroll along a sidewalk and pick out who is ovulating and who is not, Miller suggests, is because women "don't want to broadcast it" the way female primates do, with their bright rumps, because women don't want to be sexually harassed. "Instead, you want to narrow-cast these cues to high-quality guys you are interested in without leaking those cues to the boyfriend or other women, like a guy's female partner." So Susan holds back with Steady Eddie and opens up for the Cad. "It's likely there has been an evolutionary arms race, with males being under selection to detect fertility as accurately as possible, and women being under selection to send these cues only to select possible mates," Miller says.

Durante agrees with Miller but thinks men are behind in the race thanks to the ability of women to wield technological weaponry. "We can trick their brains into believing we are young and fertile, even when we are old and perimenopausal," she says with an undisguised conspiratorial chuckle. Durante, who refers to the Cads—and attractive men in general— as "Clooneys" (as in George Clooney), calls this phenomenon the Demi Moore–Ashton Kutcher effect. She thinks the wish to create a reasonable facsimile of ovulatory fertility is what drives the plastic surgery and cosmetics industries.

Sometimes the estrus-associated behavior is only indirectly aimed at attracting a high-quality mate. The more direct target is often other women competing for access to Clooneys. Women often flatly deny this, "so, when you ask them in surveys to see how women compete, you get null results," Durante says, adding, with a little sarcasm, "because we want to make sure men know we are communal and loving and warm." Women also tell her they wouldn't try to use more makeup or dress seductively. "They say, 'I would never dress sexy! I would increase my intelligence.'" But, like women's response to the question of whether they feel more lust around ovulation, this answer better reflects what they think is acceptable to themselves and to society rather than what they actually do.

Durante proved this by setting up an online fashion store. Test subjects were told that their help was needed in marketing clothing and accessories. In reality, Durante wanted to see how women's own purchasing behavior might change around ovulation. The women were told to "select the ten items you would like to own for yourself and take home with you today." Half the products were pretested to be "sexy" and half had been previously judged to be "less sexy." All items were similarly priced and carried no brand names so as not to bias choices. During the pretesting phase of the experiment, Durante had exposed other women to the site, asking them if the shopping experience made them think about attractive women or average-looking women and to compare themselves with such women. She found that the act of shopping for clothing primes women to compare themselves with more attractive women.

In the main experiment, women at their monthly fertility peaks chose sexier items than they did when they were at low fertility.

During another phase of the experiment, shoppers were shown photos of women they were told were students at the same university. Some photos featured very attractive women; other images displayed homely women. Ovulating women who had seen photos of attractive local women picked out significantly more of the sexy fashion items than did ovulators who viewed images of plain-looking women.

"The direct motivation is 'Well, who are my competitors?'" Durante explains. "'How attractive do I have to look, how hard must I work? Do I live in Malibu, where everybody looks like Jennifer Aniston, or Milwaukee?'" The more attractive a woman's direct competitors, the more pressure a woman feels to match up. "There could be a million Clooneys in the room, but going after them, dressing for them, is only effective if you have first ensured you are more attractive than other women in the local environment."

Durante's subjects weren't conscious of making different choices or aware that their behaviors were shifting. It's also important to note that her study population was mostly under forty and single. Older women and women securely mated to high-quality males don't tend to show these strong behavioral shifts. And Durante stresses these are average results on a bell curve.

Still, she's struck by the powerful effects of hormonal shifts on the brain. "Anecdotally, I see women come into the lab on low fertility days, and they're wearing their glasses, and we take their picture for the studies, and then, when they come in at ovulation, the glasses are gone; they've taken trouble to put in their contact lenses. In videos, just watching them, without looking at the chart of their cycles, we can see, wow, they are ovulating, just by looking at their faces."

All this supports the idea that human females experience estrus, that it is not hidden, and that, at ovulation, a fertile woman's brain drives her to behave in a way that will maximize her chances of mating with the fittest, and most accessible, male she can find. Men, in turn, respond with higher testosterone, which helps motivate them to engage with desirably fertile women.

When men already mated to ovulating women sense estrus, they not only want to be around the "hotness" of their mate, they want to keep other men away from it. Boyfriends engage in more "mate-guarding" behaviors when their girlfriends are ovulating; Susan's behavior in her boyfriend's absence from the lab helps us understand why. Mate-guarding men tend to interrogate their girlfriends and wives about where they're going and whom they're seeing. They snoop through their lovers' belongings; they display more jealousy. But some mate-guarding behavior is positive. Men more frequently compliment their female partners, try to spend more time with them, and express feelings of love and commitment to them when they're ovulating.

When Durante speaks about her work, she sometimes experiences a backlash from women. "I hear, 'Do you think your research is a slap in the face to women?'—and I am dumbfounded." But she says she tries not to "think about the politics of it" because the science cuts the way it cuts. Like any good researcher, she develops hypotheses to be tested, sometimes based on her own anecdotal observations. After living in Los Angeles, Austin, London, Boston, New York, and Minneapolis, Durante—who happens to be a very attractive young woman with long dark hair, dark eyes, and a lovely smile—has noticed that her own relative attractiveness waxes and wanes in relation to her "local competition." "If I am in my hometown, I'm hot," she explains with a laugh. "If I'm in Los Angeles, I'm like, 'Oh my God, my mate value is shifting!'" Then she tests these hypotheses using the scientific method. She's not making this stuff up: women and men really *do* change behaviors. One way she tries to disarm accusations of antifeminism is by asking women to pay attention the next time they near ovulation and to mentally record their thoughts, feelings, and fantasies. "Once they are aware of it," she says, "they say, 'Yeah! I had all these dreams about Jake Gyllenhaal!'"

3

THE POWER OF APPETITE

In May 2011, Jack T. Camp walked out of the Federal Correctional Institution in El Reno, Oklahoma, after serving a thirty-day sentence. A thirty-day sentence in a federal medium-security prison wasn't a bad deal, considering that Camp had been arrested on drug and firearms charges, crimes that often result in multiyear hard time. The sentence may have been lenient, but Camp suffered plenty. He was a federal judge, appointed to the bench by President Reagan in 1988, a graduate of the Citadel, an army veteran, family man, and partner in a successful law firm. His conviction wiped out a reputation built over a lifetime.

Camp was sixty-seven when FBI agents arrested him for buying amphetamines and prescription opiates from an undercover agent. He planned to share the drugs with his paramour, a twenty-seven-year-old stripper and convicted drug dealer named Sherry Ramos. Their affair began several months earlier, when Camp met Ramos in an Atlanta strip club.

Naturally, when Camp was arrested, nobody who knew him could believe the charges. After he pleaded guilty, his defense attorney filed papers with the court arguing that something—perhaps bipolar disorder or a bicycling accident years before that injured his temporal lobe—must have impaired Camp's ability to control his impulses. While not an excuse, the filing stated, such an accident did "help explain . . . how in May of 2010 a lonely man in the twilight of his life became entangled with a seductive prostitute." The statement was widely ridiculed.

Five years before Camp's arrest, Abraham Alexander's employers at the Cardiovascular Research Foundation, in New York City, were shocked to discover that their trusted friend and employee had embezzled nearly a quarter of a million dollars from the nonprofit. Alexander was the foundation's accountant. A happily married man by all appearances, he had a nice home in East Meadow, on Long Island. "He was a quiet little guy," his lawyer, Hershel Katz, tells us.

Alexander had no criminal record, no hint of trouble, in his past. Yet he pleaded guilty to stealing the foundation's funds, which he used to pay for travel, accommodations, and services on frequent trips to Columbus, Ohio, where he visited Lady Sage, a professional dominatrix.

Such stories may remind you of the US senator and the New York governor who both visited prostitutes; the pretty female teachers who have seduced teenage boys in their classes; the married California governor and movie star who fathered his maid's child; the conservative preachers; the US presidents, French politicians, and British MPs brought low by sex scandals of one stripe or another. Most sex scandals, though, are neither felonious nor public. They take place every day, among millions of people all over the world: the wife who cheats, the boyfriend who falls into bed with the girl he just met, the teenager who swore she'd remain a virgin until marriage. Each one represents a failure to conform to society's rules and expectations, or simply to our own, often at the expense of our self-interest.

The story of Abelard and Heloise may be one of history's most romantic love stories, but it was a sex scandal. Nearly 900 years before Judge Camp's arrest, Abelard, a young French ecclesiastical scholar with a reputation for "continence," found himself lusting for a younger woman, the comely Heloise. He knew, as he later wrote, that any carnal contact between them was "directly contrary to Christian morality" and "odious to Jesus Christ." Yet, as Abelard recounted in his appropriately titled apologia, *Historia Calamitatum*, he couldn't help himself. Under the influence of rising testosterone and activated heterosexual male circuitry, he connived to trick the girl's uncle into appointing him her tutor. Not only did the two become lovers, but they also appear to have explored what we now call sadomasochism.

> Under the pretext of study we spent our hours in the happiness of love, and learning held out to us the secret opportunities that our passion craved. Our speech was more of love than of the books which lay open before us; our kisses far outnumbered our reasoned words. Our hands sought less the book than each other's bosoms— love drew our eyes together far more than the lesson drew them to the pages of our text. In order that there might be no suspicion, there were, indeed, sometimes blows, but love gave them, not anger; they were the marks, not of wrath, but of a tenderness surpassing the most fragrant balm in sweetness. What followed? No degree in love's progress was left untried by our passion, and if love itself could imagine any wonder as yet unknown, we discovered it. And our inexperience of such delights made us all the more ardent in our pursuit of them, so that our thirst for one another was still unquenched.

Now that they had experienced sex, retreat from their forbidden lust was impossible. Their passion became so powerful that Abelard, like Judge Camp and Abraham Alexander, began behaving in very uncharacteristic fashion. He stopped studying. Just going to the school became "loathsome." He frequently skipped school altogether.

As will happen when two young, fertile people have lots of sex, Heloise became pregnant. The uncle discovered the affair and agreed to allow a marriage, but then he double-crossed the couple by having Abelard attacked and castrated: "For they cut off those parts of my body with which I had done that which was the cause of their sorrow. . . . I saw, too, how justly God had punished me in that very part of my body whereby I had sinned."

Abelard then lived the life of a monk; Heloise, a nun. Abelard, probably because he was missing his gonads, seemed much better able than Heloise to adapt. "Among the heroic supporters of the Cross I am the slave of a human desire," she wrote to Abelard. "How difficult it is to fight for duty against inclination. . . . My passions there are in rebellion; I preside over others but cannot rule myself."

Though many of us can identify with Heloise's internal conflict, Dante reserved the second level of hell for such people.

And this, I learned, was the never ending flight
Of those who sinned in the flesh, the carnal and lusty
Who betrayed reason to their appetite.

Yet the appetite to which Dante refers is an important driver—not only of our economy but of our love. It's the very same appetite so often piqued by cosmetics companies and beer brewers and tool makers, by the Pirelli tire company that publishes famously sexy calendars. Business empires have been built on it. Back in 1953, Hugh Hefner was a skinny, bookish editor, formerly with *Esquire*, who sat down at a table in his home and pasted together what became the first issue of *Playboy*. His creation was not entirely original; it was a combination of two existing forms: the sophisticated, urbane text of *Esquire* and the lowbrow, pulchritudinous imagery of cheaply produced, under-the-counter nudie magazines. But by the 1970s, *Playboy* had a monthly circulation of over six million, and Hefner was flying around the world in a custom DC-9 painted with the famed bunny-head logo.

Such manifestations of our lusty appetites are often in conflict with thousands of years of laws and rules, moral teachings, manners, and self-imposed restrictions—all designed to curb passions. Yet the same appetites that made Hefner rich are the ones that lead us into the passionate love we extol as one of mankind's highest ideals. The resulting friction is the stuff of which literature is made, but both the glory and the shame derive from the same brain circuitry.

The brain battle between "want" and "should" is waged in many aspects of life, not just the erotic. Here's a practical, nonsexual example that illustrates a problem most of us face: Larry is passionate about his mother's cooking, a cuisine rooted in rural Georgia. He was weaned on fried chicken; salty, fried okra; buttery grits and mashed potatoes; sweet tea; the occasional barbecued goat. Larry makes delicious versions of all these dishes. But the first time he cooked for Brian, Brian drove away from Larry's home with one finger poised over the 911 button on his cell phone in case of cardiac arrest. Now let's say that, like us, you find these foods temptingly tasty. Unfortunately, your cardiologist disapproves, so you try to eat only salads, egg-white omelets, and steamed fish. You may not like it, but you tell yourself you'll be able to stick to the diet because you've made the reasoned calculation that you'd rather live into your nineties than go down grabbing your chest in your fifties. While you love fatty, salty vittles, there's not a Lipitor big enough to ream all that cholesterol out of your arteries.

Because you're disciplined and rational, you manage to stick to the diet. But one day (probably sooner than later), you find yourself halfway through a groaning platter of crispy double-battered, deep-fried chicken, butter-mashed potatoes with salted giblet gravy, and fried hush puppies—because, come on, how can you not have the hush puppies?

Your rational self tried to conjure up a vision of you lying on an emergency-room gurney as doctors pressed those electrified paddles to your chest and shouted, "Clear!" But that guy was overwhelmed by your appetite. It's the old "devil" Daffy Duck versus "angel" Daffy Duck perched on opposite shoulders routine. Devil Daffy not only has a megaphone, he can also outsmart Angel Daffy with cunning rationalizations ("I'll do an hour on the elliptical trainer!").

Behavioral economist George Loewenstein refers to this as a "hot/cold empathy gap." In the car coming home from your cardiologist appointment, you were in a bloodless "cold" state. Using your rational mind, you dedicated yourself to poached fish and greens with the zeal of a convert. Later, when you experienced the influence of what Loewenstein calls "visceral factors," you entered a "hot" state. You underestimated the difficulty you would have resisting fatty foods—the empathy "gap" within your brain—and you quickly caved.

This phenomenon has been explored over the past twenty years by researchers such as Loewenstein and Dan Ariely, whose 2008 book, *Predictably Irrational*, was a best seller. While terms like "hot/cold empathy gap" and "visceral factors" may accurately describe phenomena, however, they don't explain the biology driving them. As we saw in the last chapter, it is this drive, working on activated female brain circuits laid down during fetal development and after, that influenced Susan's behavior during her conversation with the Cad. This was a beginning, a nudge, as her brain became biased toward mating when she encountered a desirable candidate. But she didn't leave with the Cad. While the bias raised her appetite enough to alter her behavior, she didn't become the Heloise of Minnesota. Something much more powerful drove Abelard and Heloise, Camp, and Alexander.

THE CASE OF THE HEDONISTIC RODENTS

You might regret learning this, but it could come in handy someday for settling a bar bet: Not only is it possible to masturbate a female rat, but when you do so, she'll make happy little squeals that sound like a baby bird—*eep, eep, eep.*

Two female albino rats are making *eep, eep, eep* noises in the basement lab of a building on the campus of Concordia University, in Montreal. Two young women, Mayte Parada and Nicole Smith, dressed in white lab coats and latex gloves, are picking up one rat each, and giving the rodents' clitorises five quick strokes with small, soft artist's paintbrushes that have been dipped in sex lubricant. Then they set the rats down on the flat top of a small cart. An assistant counts off five seconds, and Parada and Smith repeat the process. After the first few repetitions, the rats don't want to be released back onto the cart. When Parada and Smith set them down, the rats look back up at the women with disappointed eyes. After the fourth set of strokes, Parada's rat clings to her arm, scrambling up past her wrist and tucking itself into the crook of her elbow.

"I love you, I love you," Parada says, in her imagined rat's voice. The scene looks like it could be from one of those esoteric porn videos featuring fetishes you never knew existed.

If you were presented with a police lineup of neuroscientists and asked to pick the guy whose lab masturbates rats, you'd pick Jim Pfaus. With a pair of rings in one earlobe, one of those spiky-thorny tattoos encircling his upper right arm, and a dark Vandyke that gives him a touch of the satanic, Pfaus could easily be mistaken as a 1980s adult-entertainment impresario surrounded by big-haired, nubile women in purple Lycra jumpsuits. He is the very rare neuroscientist who can quote from *Monty Python*, Pavlov, *Deep Throat*, William James, Suzanne Sommers, Stendhal, and punk rocker Jello Biafra in the space of ten minutes. As an undergraduate at American University, a Mohawked Pfaus became part of Washington, D.C.'s late-'70s–early-'80s punk scene, singing and playing guitar in the band Social Suicide. Pfaus went on to graduate school at the University of British Columbia, then worked with Don Pfaff before landing at Concordia. When he wasn't fronting a new band called Mold, Pfaus was studying the brain mechanisms that make sex feel good—and how that good feeling leads to behavior.

Like most scientists, Pfaus wanted to know the answers to these questions because he was curious. Unlike most scientists, he can recall the exact moment when his curiosity became an obsession. The son of a government labor official and a music teacher, Pfaus grew up as an intellectual lost boy who analyzed just about everything. So, when he had his first self-induced orgasm, he wasn't satisfied merely thinking to himself, "Cool!" or "Did I break it?" as most boys do. Instead, he tried to diagnose it. "I was, like, 'My body never did *that* before. What was going on?'" he recalls.

Similarly, when Pfaus got to know junkies in D.C., he was neither repelled nor tempted to join them. Rather, he says, "I wondered, 'Why are these people pursuing coke or meth or heroin?' And they would describe the feeling, and it sounded like sex, and I thought, 'Now I know what *The Iliad* is about! It's about sex!'"

The leap from junkies to sex to *The Iliad* is typical Pfaus, but as disconnected as these themes may seem, they constitute a pretty accurate summation of what happens in the human brain. And, as you'll see, Larry is convinced that the mechanisms of irrational passion Pfaus and others explore are pillars in the construction of human love.

Pfaus's scientific story really begins in 1953, the same year as *Playboy*'s founding, when a new Harvard PhD named James Olds began his post-doctoral research at Montreal's McGill University. In his first experiment, Olds tripped over the discovery of a lifetime.

Electrophysiologists enjoyed a heyday in the 1950s and '60s, when scientists found ways to artificially trigger behavior with electricity (a first step in mapping the brain). To do so, they implanted electrodes in the brains of animals.

Olds wanted to study what was called the reticular system, the term given to a somewhat amorphous group of neurons deep in the midbrain that acts as a gatekeeper, telling the brain to pay attention to some sensory inputs and ignore others. So Olds placed electrodes into what he hoped was the reticular system. As he later admitted, he was new at this sort of thing and had bad aim. Not all the electrodes hit the target in every animal.

His experiment was simple. Olds would let a rat wander around an open arena as he used a control button to periodically send a small zap of electricity into the animal's head. Then he observed whether the jolt changed behavior in some way. His first observations were mundane. But when one rat began creeping across the floor of the arena, and Olds pushed the button, the animal instantly stopped, backtracked a couple of steps, and looked up at the startled scientist. "The rat seemed to say, 'I don't know what I just did, but, whatever it was, I want to do it again,'" Olds later told colleagues.

As the testing went on, Olds found he could cause the rat to favor a particular corner of the arena by jolting it when it visited that corner. If he stopped administering the shocks, the animal would lose interest and wander again. Olds could then direct it to another corner.

At first, Olds thought he was simply sparking curiosity. Then he set up a walkway ending in a T intersection that branched into right and left arms, sort of like a fashion-show catwalk. He discovered the rat would turn into the arm that led to a brain buzz. Next, he put the rat on a twenty-four-hour fast. After the fast, he put food at both ends of the T arms and placed the rat at the start of the walkway. If left alone, any normal, hungry rat, smelling and seeing the food, will dash to one of the piles of chow and gobble it up. But when Olds gave this rat a hit of electricity as it walked down the straight path toward the intersection, it stopped, losing all interest in food. The rat liked whatever was happening in its brain much more than it wanted to eat.

Olds and Peter Milner, a McGill colleague, then set up a new experiment. This time, they put electrodes in rat brains at a variety of regions, including the one Olds thought he'd hit in the remarkable rat. They placed the animals, one at a time, in a Skinner box (named for renowned behaviorist B. F. Skinner). The box was equipped with a lever that when pressed would release a hit of electricity into the brains of the rats. Each time the scientists dropped a rat into the box, they tapped the lever to show the rat how the lever worked. Then they left the animal alone to do what it wanted.

Some of the rats avoided the lever. Others loved it. Rat number A-5 pressed the lever 1,920 times per hour, or once every two seconds. Olds and Milner didn't know it yet, but the electrodes in A-5's brain had plumbed the reward system, a collection of interconnected areas including the ventral tegmental area (VTA, where dopamine is made), the medial forebrain bundle (it connects the VTA to other brain areas), the septum, the hypothalamus, and the amygdala. (See Figure 2, page 76.)

Once they dissected poor A-5, Olds and Milner realized they'd found the circuitry that creates the good feelings when an appetite, like a craving for food or sex, is appeased. More than that, though, they discovered that this system can actually control behavior to the point of destruction.

The first hungry rat stopped short of the food when he felt the pleasurable zap in his head: a decision that would seem directly opposed to his self-interest. When Olds and Milner placed electrodes in just the right spot and left the rats to their own choices, the animals would keep pressing the lever—ignoring food, water, sleep—pressing and pressing, sacrificing all other concerns until they pleasured themselves to death.

Many scientists in the post–World War II decades hoped to perform electrophysiology experiments on people, but it's tough to recruit volunteers willing to have metal electrodes shoved into their brains, and, even if they could have done so, academic administrators were understandably squeamish about the idea. This state of affairs frustrated psychiatrist Robert Galbraith Heath. Despite holding a prestigious position at Columbia University in New York, where he studied schizophrenia, Heath chafed under the university's ethical restrictions. He could experiment on rodents, and sometimes monkeys, but he wanted access to human beings.

Tulane University took a different view from that of Columbia. The school had big aspirations of becoming a major intellectual hub in the South, but it had trouble attracting top-flight talent. When the university's ambitious pooh-bahs decided they wanted to start a psychiatry department, they targeted Heath to be the man in charge. Compared with Columbia, Tulane was a backwater. But when the school looked around New Orleans and the state, it realized that the city's large Charity Hospital, which served the poor, and the state of Louisiana's mental hospitals represented a deep pool of possible human experimental subjects. Tulane offered Heath access to this vast amount of what he called "clinical material," and he joined the faculty in 1949.

The next year he began placing electrodes—sometimes more than a dozen at a time—into the brains of people. He often noticed that shooting electricity into certain brain regions produced pleasurable feelings, much like what Olds and Milner would later find in rats. Unlike rats, though, people can talk. When they described the nature of the pleasure they felt, they sometimes told Heath it was most definitely erotic.

In 1972, Heath conducted an especially notorious experiment (in a career of notorious work), during which he tried to convert "B-19," a twenty-four-year-old gay man, into a heterosexual by implanting eight electrodes into his septal region. Heath combined the sensory reward of the zaps with a pornographic movie, and the attentions of a twenty-one-year-old prostitute, so that B-19 would associate pleasure in his brain with heterosexuality. Eleven months after the "therapy," Heath declared the experiment a triumph and suggested using brain stimulation as a way to reinforce desired behavior and "extinguish" undesired behavior (thus providing antipsychiatry fodder to Scientologists and mind-control conspiracy theorists everywhere).

In fact, B-19's "conversion" should be viewed with great skepticism. Heath, who was not in the room during the tryst between the prostitute and B-19, relied on her version of what happened. She claimed a great success, with orgasms all around, despite the fact that B-19 had never before had sex with a woman, that the wires sticking out of his head and connecting him to a machine made the gymnastics of sex a little awkward, and that prostitutes rarely climax with their clients. Even so, B-19's post-therapy story of conducting a brief sexual relationship with a married woman (not interviewed by Heath) who never allowed B-19 to ejaculate in her vagina, and his assertion that he'd had gay sex "only" twice since his treatment, were enough for Heath to declare victory over homosexuality.

While Heath most likely didn't change B-19 into a heterosexual, he did note two important observations. When he juiced B-19's brain, the man, who had long used drugs, said it felt a lot like using amphetamines. And when Heath allowed B-19 to control his own stimulation, the scientist soon had to take the controller away: B-19 kept hitting the trigger over and over, like Olds and Milner's rats.

Erotic pleasure in the brain can lead to this kind of obsessive behavior. In 1986, doctors reported the case of a woman who'd had electrodes implanted deep in her brain in an effort to treat the excruciating pain she endured as the result of a back injury. The brain stimulation did help relieve her pain, but it also created intense erotic pleasure (though without orgasms). She became so enraptured by the sensations, she rubbed sores on her thumb from manipulating the dial on the controller. She would sometimes spend entire days stimulating herself, neglecting time with her family, her own hygiene, even eating. At one point she gave the controller to a family member, with orders not to return it. Then she begged to have it back.

MORE LIKE THAT

Work in Pfaus's lab, and in those of others', is now showing that physical sexual stimulation can have the same effect. Parada is stroking rat clitorises with lube-dipped paintbrushes to find out if intermittent, as opposed to continuous, stimulation of the clitoris can lead female rats to adopt a "place preference," and if variations in the rats' brain chemistry factor into whether they develop such a preference. Experiments over many years, in many labs, have shown that rodents like to be in places where good things happen. Feed a rat a few times in one chamber, then give it a choice to be in that chamber or in an adjoining one, and, most of the time, rats will choose to loiter in the chamber were they've been fed. Let a rat have sex in a chamber with, say, an uncomfortable grate on the floor, and then give it the choice of hanging out in that chamber or in one with cushy wood shavings on the floor, and the rat will most often choose the uncomfortable room.

Parada is using a light/dark choice. Rats naturally prefer dim spaces. But Parada is discovering that if she keeps a rat waiting for the stroking sessions in a chamber painted with silvery metallic paint, then returns it to a connected chamber with dark walls, the rats will move to the bright chamber.

Parada's intermittent clitoral stimulation mimics what animal behaviorists call "pacing." If female rodents are allowed to, they'll control the pacing of sex to their own liking. When a female is placed in a two-chambered cage with a divider she can cross, but that a male cannot, she can decide when to let the male mount and when to keep him in his place. If she's in estrus, she'll cross into his chamber, hopping and darting to seduce him, then adopting the lordosis position. After a few intromissions, she'll dash through the opening and back into her chamber before returning for more, creating just the rhythm of sex she wants until she's satisfied. Later, if she's put into that enclosure again, even with no male present, she'll spend most of her time in the chamber that housed the male she used for paced sex.

In other words, the rodents remember where they experienced pleasure, and they associate that pleasure with the environment they happened to be in when it came. That experience is so powerful it can make animals override their natural aversions, like being in a bright space.

It can also cement partner preferences. If a female is allowed to control the pacing of sex with a male who has been artificially scented, or who is pigmented differently than other males so she can pick up cues associated with the pleasure, she'll often prefer to have sex with that male over other males. She won't be monogamous, but she'll play favorites.

Rat polygamy is a fixed, innate behavior. But Pfaus has recently begun to wonder if even *that* can be modified by sexual experience.

When a female develops a preference for a male, she does so mainly on the basis of scent. Which is why, for years now, Pfaus's lab has experimented with scents and mating behavior. (Pfaus uses artificial scents because males are lousy at identifying individual females based on natural ones; females are better at using natural scent to discriminate among males, but for some experiments they also need the help of artificial smells.) For example, if male rats have their first sexual experience with a female who has been daubed with a lemon scent, the males, when given a choice between the lemony female and a natural-smelling one, will choose the lemony female. This is true, even though the smell of a female rat in heat is very attractive to male rats and even though males usually like to mate with many different females.

Scientists in Pfaus's lab wondered what might happen if they allowed the normally polygamous rats to copulate with only one partner—a partner that was not scented with any artificial aroma, just the normal smells of female rats. They gave the couple time to mate repeatedly, anticipating that the male would form a preference for his girlfriend. But they were surprised to discover that when they added a second female to the couple to create a ménage à trois, the males were equal opportunity ejaculators: they didn't prefer either female. They were also surprised by what the males' putative girlfriends did: They became jealous, actively trying to thwart the males from mating with the female intruder. Rats aren't supposed to do that, yet the girlfriends turned aggressive toward the second female and tried to block the male's path to her.

Because the females weren't the object of the experiment, the scientists ran it again. This time, though, they daubed an almond scent on the would-be girlfriends. Unlike the first test, the males, now with the help of a strong olfactory cue, showed a marked preference for ejaculating only with the girlfriend during the ménage à trois, though they would still mount the intruder. This time, the girlfriends became even more violent toward the second female.

"When she wore the odor, she was hyperaggressive," Pfaus recalls. "She would beat the shit out of the other female. I cannot give you anything other than an anthropomorphic explanation, which is, 'If I am wearing my sexy odor, he damn well better do me, and only me. She has associated the wearing of the odor with him and the expectation that she will have sex only with him. She does not know what we know, that he will direct his ejaculation only to her; all she knows is that he is screwing the other female. So she beats her up and solicits him repeatedly. She becomes the major party girl."

Such deviations from natural rodent behavior can be traced directly to the same brain-reward system Olds and Milner discovered. While Pfaus and others work mainly with rodents, and while rodent experiments are not automatically transferrable to people, anecdotal evidence has long indicated that humans also develop a preference for places in which they've had rewarding sex, even if those places—cheap motels, Las Vegas, 1982 Chrysler LeBarons—may not seem all that appealing.

In 2010, University of Chicago researchers confirmed what many had already concluded by using a sex stand-in: amphetamines. They found that people given small doses of speed developed a place preference for the room in which the dose was given, while those who'd received a placebo didn't develop a place preference.

The effects of reward go beyond simply preferring one place or one sex partner. Animals, including people, like sexual reward so much they'll work for it. Barry Everitt, of Cambridge University, proved this when he rigged up a rat pleasure dome. In part of the experiment, he created a place preference in male rats by providing them with a willing female. But he added a twist. On one wall of the cage, he installed a small, circular light that was turned on whenever the rats mated.

Next, Everitt put the males back in the box, now equipped with a lever that would turn on the light. The males quickly learned to push the lever to turn on the light. They didn't get a food treat or a female—at least not right away. The light itself produced a brain reward because it had come to be seen as a cue for sex. About fifteen minutes after a male had switched on the light, a female in heat would drop out of the ceiling like candy dropping from a piñata.

This is called reward-based learning. And while willing women won't drop out of ceilings when men turn on light switches, men and women have learned to work for sex. Or, for those of us who think life should be more like *Top Hat*, we'll act romantically. We'll flirt. We'll buy flowers, even though we don't know a gardenia from an orchid. We'll wear sexy clothes, even though we're more the jeans-and-sneakers type. We'll find ourselves agreeing that we can totally taste the difference between *saku* cut tuna and whatever isn't *saku* cut tuna even though we have no idea what you are talking about, and every time we look at those little slabs of rice with raw fish on top, we'd kill for a decent rib eye.

If we were a rat in Everitt's experiment, we'd be turning on the light.

"We do not have to groom ourselves, we really don't," Pfaus explains. "This excessive grooming of hair does not change your appearance one iota. Yet, we do it. We do not have to wear our lucky socks on a date, but one time, when we wore them, we got laid, so we wear them the next time. . . . This arises because of the brain mechanisms Everitt was exploiting. The rat in Everitt's experiment knows the light represents sex."

Put another way, the light is like buying sexy clothes for the women in Durante's studies, a trigger for appetitive reward even without actual sex. The same with money. Obtaining money produces brain reward on its own, but one of the reasons it produces brain reward is that it's so often associated with sex. Men with more money get more sex, or at least more opportunities for sex. And not just sex, but sex with a better choice of partners. "Ask any guy, and he will tell you that when you see a hot chick, you gauge how much work you are going to have to do to seduce her," Pfaus says.

As most women know, once males ejaculate, we'll stop working for sex, which accounts for the rapid turnaround from "I'm taking you to Paris for your birthday," in the middle of foreplay, to "There's no way we can afford it," after ejaculation. It is the wanting that drives the Armani wearing, the hair-gel smearing, the cash flashing. This is appetitive behavior—like the female rodents' hopping and darting, like Susan's flirting—and we receive a brain reward when we do it.

Desire can start internally, when our brains are primed with estrogens (as in the case of ovulation in women), or androgens in men, and reach out into the world (as in "I'm horny and I want to have sex"). It can also come from the world and into our brains if our genitals are stimulated, or if we encounter a sex-related environmental cue that could involve just about anything—from a man in uniform to the lingerie rack in a department store.

A general state of arousal helps. We aren't necessarily referring to being sexually aroused, but to arousal of the sympathetic nervous system. If you've ever taken a bungee jump or dived out of an airplane, you've probably experienced a feeling of exhilaration that lasted hours, even days. Jumping off a bridge after being secured to an elasticized cord by a teenager earning minimum wage because he flunked physics leads to a big shot of noradrenaline (also called norepinephrine). Your heart beats faster, your mouth becomes dry, you're hyperalert—ready for fight or flight. But once you realize you aren't going to die, you can feel almost euphoric.

There's really no need to jump off a bridge, though. A good comedian, a couple of espressos, exercise, even a spanking can all have the same effect: They generate arousal through laughter, caffeine, exertion, or slight pain. Novel experiences can work, too. Even long-married and sexually bored couples can suddenly find themselves groping each other over breakfast when they take a vacation. They're eating different foods, meeting different people, walking through unfamiliar streets—all of which can induce a low-level, even pleasant, anxiety.

This is akin to what Pfaus calls the "dud-to-stud" phenomenon. "Low-to-moderate levels of shock and pain can make sexually sluggish or inactive male rats turn into semivigorous copulators," he explains. Activation of the general arousal system is neutral; it doesn't have to lead to sexual desire. If you're surrounded by food, you might start eating, even if you aren't hungry. If you're surrounded by sex cues, you'll likely be interested in sex. Arousal is "given 'meaning' by whatever is in the environment at the time," Pfaus explains.

This "meaning" comes from what Pfaus calls the "bottom" of our brains—the hypothalamus and the limbic system. For humans or rats or monkeys, appetitive behavior is based in this system, which consists of the medial preoptic area (MPOA), the nucleus accumbens, the amygdala, and the VTA. Our appetites are aroused, and we're motivated to act by neurochemicals binding to receptors on neurons in these regions.

Borrowing slightly from William James, Pfaus likes to refer to the interplay between our brains' executive functions and the functions of the limbic system as "top-down versus bottom-up" or, as Loewenstein might say, cold versus hot. The "top" of our brains, thinking of competing interests, are always weighing and judging and making this-versus-that calculations—often saying no. The double-battered, deep-fried chicken like Larry's mother used to make appeals to the limbic system, but the worry about keeling over comes from the top.

As Olds, Milner, Everitt, and years of work in Pfaus's lab and elsewhere have proved, reward can tip the balance of power from the top to the bottom. The tipping point rests largely in the hypothalamus, the

MPOA in particular, which acts something like a traffic cop. In addition to functions like regulating temperature, directing blood flow, and controlling hormone secretion, the MPOA chooses from among environmental cues to appease appetites like thirst, food, and sex. If you have never attempted to mate with a boiled lobster, you can thank your MPOA.

Several events occur at virtually the same time to create appetitive reward. First, steroid hormones trigger the synthesis and release of alpha melanocyte-stimulating hormone (MSH), which, in turn, drives the release of dopamine in the MPOA. The MPOA strongly signals the VTA, which sets off limbic system structures like the accumbens. Glutamate originating in the amygdala also tells dopaminergic neurons in the hypothalamus to release dopamine in the MPOA. This dopamine binds with one of several types of dopamine receptors, including D1, sometimes referred to as the "short-term-desire" receptor. Meanwhile, a small amount of opioids, the brain's version of heroin, is released. That creates a pleasant feeling.

When all that dopamine hits D1 receptors on MPOA neurons, we become very attentive to sex-related cues. (If we were starving, the arousal of our hunger, combined with dopamine in the MPOA, would focus our attention on food cues.) Second, the MPOA directs the parasympathetic nervous system to send blood to the genitals, creating erections in males and clitoral engorgement in females. (Exactly how this works is not yet known, but it involves another hypothalamic structure, the paraventricular nucleus, which is connected by neurons to the MPOA. Sexual cues also stimulate the release of oxytocin, about which you will read much more, and oxytocin and dopamine in this nucleus help create lordosis in female rodents and erections in male rodents.) Third, after receiving signals from the MPOA via neural projections, the VTA transmits dopamine into the prefrontal cortex, part of our executive brain. (See Figure 2, page 76.)

Dante may have condemned people who betray their reason in favor of lust to the first circle of hell, but he didn't know that nature—or God, if you prefer—designed this betrayal into our brains. Dopamine mutes the prefrontal cortex, disinhibiting sexual desire and giving us tunnel vision

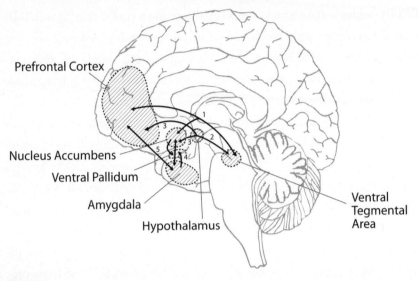

Figure 2.
Neural connections involved in sex, love, and attraction. (1) Dopamine activates reward system. (2) Oxytocin (OT) promotes dopamine release in new mothers. (3) OT is released into reward systems. (4) Vasopressin from the amygdala stimulates reward systems. (5) The prefrontal cortex applies brakes to subcortical drives. Each of these circuits is discussed in relation to sex, love, and infidelity.

for cues that could lead to satisfying that desire. When young men look at pictures of naked women, their startle response to loud noises becomes blunted. When ovulating women look at pictures of naked men, their pupils dilate and they smile unconsciously, much as Susan reacted to the Cad.

Of course, booze and drugs can mute the prefrontal cortex, too. The parental "top," partying with martinis in the living room, is too woozy to supervise the unruly kids playing spin the bottle in the basement. Cocaine and amphetamines can greatly enhance sexual motivation because they stimulate the release of large amounts of dopamine. Simply doing a lot of mental work can loosen the reins. Executive thinking involves comparatively massive brain real estate—the human prefrontal cortex is about ten times larger than the hypothalamus—and consumes much more energy. Loewenstein has found that doing math problems before being offered cookies can lower a dieter's resistance to the reward of a good Toll House: all that thinking has drained our executive brain's batteries.

As Augustine of Hippo told early Christians, in *City of God*, "the soul is therein neither master of itself, so as not to lust at all, nor of the body, so as to keep the members under the control of the will."

LOVE'S BAIT AND LEATHER DUDES

You might think it sad that we have to be worn out, or boozed up, or loaded with dopamine in order to have sex. But if it weren't for the reward, why would we do it? As Schopenhauer wrote, "Let one only imagine that the generative act was neither a want nor accompanied by extreme pleasure, but an affair of pure rational reflection: could then the human race continue to exist? Would not rather every one have so much compassion for the coming generation as to prefer to spare it the burden of existence?"

His fellow German, Eduard von Hartmann, argued that we have to be bribed into sex, our reason shut down, because sex leads to no good: marriage, the pain of childbirth, disappearing money, disillusionment in love. Just about the worst thing that can happen, Hartmann thought, is to become conscious that "the dreamed-of bliss in the arms of the beloved one is nothing but the deceptive bait" used to get us to procreate. We have the illusion of control, but, of course, it's our unconscious breeding instinct, working through our brains, that's driving our behavior.

According to Hartmann, the clash of biology with social norms could call for a drastic truce. "If love is once recognized as evil, and yet must be chosen as the *less* of two evils as long as the impulse persists, reason necessarily demands a third, namely, eradication of the impulse, i.e., emasculation, if thereby an eradication of the impulse be attainable." He then approvingly quoted from Matthew 19 a passage about men who made themselves eunuchs in order to reach heaven. (And people wonder why there are few good German comedians.)

This view may seem overly gloomy. But it's true that the promise of the reward keeps us from thinking too much about the possible downside of sex and makes us willing to work for it. When Everitt disconnected the rats' amygdalas, they stopped working for sex. They would no longer turn on the light because they had no appetite for sex. If he disconnected the MPOA, they'd turn on the light, the female would drop out of the ceiling, and the males would act interested, but then wouldn't consummate the deal.

Before our first orgasm, there's no specific consummatory goal, only appetite. When a little boy discovers that touching his penis feels good, or a little girl has a eureka moment with the water streaming from a bathtub faucet, they aren't usually thinking about sex—at least not sex as it will someday take shape in their brains. The sensations are their own reward. The little girl may come to relish bath time, and the little boy may look forward to spending time in the private place he associates with the pleasure of touching himself. Baths and a bedroom closet have become linked to the pleasure and can create appetitive reward like the light in Everitt's cage, though without any endgame. When researchers in Pfaus's lab stroke a virgin female rat's clitoris in a chamber scented with lemon odor, the lemon scent itself becomes rewarding to the rat and makes her desire the stroking.

But, eventually, we desire consummation. When we do, our appetitive behavior will lead to consummatory behavior. The differences in brain structure Dick Swaab and Charles Roselli are exploring relate to how, and toward whom, our appetitive desire and consummatory goal are directed. These are innate drives. Even before male rats have had any sexual experience at all, the natural scent of an estrous female triggers the release of dopamine in the accumbens of the males' brains. Before puberty, heterosexual boys and girls may insist on the general ickiness of the other gender. But despite their protestations, they're interested, even though the interest is unfocused.

Someday, say, if a boy finds himself alone with the two-dimensional loveliness of Miss October, the shape and aspect of Miss October—especially her breasts, her face, and her eyes, the three locations on a naked woman's body where studies show human males spend the most time looking—will activate his innate appetitive drive, and he'll get an erection. If he happens to be alone, he'll respond to this reward by chasing more reward and touching his penis. Miss October and others like her, including actual live, three-dimensional females, now become erotically rewarding in their own right. Miss October has become an "antecedent condition." Merely looking at her image delivers a brain reward like the light in the rats' cage. The pleasure of genital stimulation now possesses a sexual context.

Then, one day, pursuing that appetitive reward leads to an orgasm. A boy may not even be old enough to ejaculate, but from that moment on, he has a goal. He'll never again be satisfied with the appetite alone; he wants the reward of consummation.

The same process happens in females, two- and four-legged, especially with the coming of puberty. "When our females get the clit stim associated with an odor, they think it's fabulous," Pfaus says. "But if we put the male in with her, and give her the clit stim, so the real thing is there, well, now the clit stim becomes the antecedent condition and drives them to go solicit the male. That's intrinsic. They are saying"—here Pfaus raises his voice to a female-rat octave and waves his hands—"'Oh my God, it's the real thing! That clit stim gets me all gooshy, but my clit, stimulated by a paintbrush? Thanks very much for that—it was great—but I'm ovulating. I want to get fucked!'"

People and animals don't want anything to stand in their way when they've become sex-goal directed. D1 binding in the MPOA of rodents produces the female solicitation of males. Males will solve a maze, hoping to find a willing female at the end of the trail. Humans will seduce, give strippers outrageous tips, and betray their own values.

Social behavioralists like Loewenstein have conducted experiments showing just how determined men can become while in thrall to reward seeking. In 1996, he found that men looking at images from *Playboy* were more likely to say they'd try to coax a woman out of her clothes, even after she'd said no to sex, than were men who had looked at the same images a day before answering.

Ten years later, Loewenstein and Dan Ariely conducted two surveys of men, one while they were in a neutral (cold) state and the other while they were actively masturbating to erotic images on laptop computers (thoughtfully wrapped in plastic). Men in the first survey answered a series of questions about their sexual ethics in ways we'd hope for. Very few said they'd get a date drunk in order to have sex with her, for example. When asked about what they found sexually exciting, their answers were again predictable. Few said they'd be turned on by an obese woman, by shoes, by sex with animals. But when they answered the questions while aroused, significantly more men said they would indeed get a date drunk. More of them found obese women, bestiality, shoes, and a three-way with another man and a woman to be sexually exciting. In the context of serious arousal, any means to an end will be considered.

As the opioids and dopamine do their work, our inhibition falls away. "You no longer have this clash of the bottom-up, top-down, titans," Pfaus explains. "Bottom-up wins."

When Pfaus refers to this appetitive reward as "cocaine-like," he's not just using an analogy. Appetitive sexual reward is *exactly* like appetitive cocaine and methamphetamine reward. Infusing amphetamines, which prompts the release of dopamine, directly into the nucleus accumbens of male rats makes them extraordinarily eager to have sex. That's why B-19 told Heath the electricity from the electrodes gave him an erotic sensation that felt like taking amphetamines. When cocaine and methamphetamine users are placed in fMRI machines and shown neutral pictures, sexy pictures, and pictures of drug paraphernalia like straws and razor blades and piles of white powder, their limbic system activates in the same manner to both the sexy pictures and the drug-related ones.

This is how Hugh Hefner got rich. Without understanding exactly what he was doing, he tapped into an innate desire to look at female breasts and faces—causing millions of males all over the world to experience appetitive reward, to plunk down their cash for the magazine, or to look forward to its delivery every month. *Playboy* itself became an antecedent condition leading to consummatory reward for its readers, and an L.A. mansion and DC-9 jet for Hefner.

This need for consummatory reward is so great that if animals are foiled, their preferences can be broken. When Pfaus puts an estrous female in an enclosure with a male that's placed behind a screen so he can't touch her, the male's scent "provides odor and pheromonal cues that are unconditionally appetitive," Pfaus says. This "activates her amygdala, her medial preoptic area, gets her ready to solicit, but, in our condition, she was teased. She was all dressed up, with nowhere to go. She solicits the male behind the screen, but he can't do anything. So, as far as she is concerned, he's a big tease." As a result, even if she's later given access to that male, if another male is present, she'll avoid the tease—or cues associated with the tease and the subsequent frustration she experienced—and mate with the male who can provide some follow-through.

Nobody knows if female rodents, or even male rodents, have orgasms, though Pfaus likes to think they do. But whether they do or don't, they receive a consummatory brain reward, just like humans. Dopamine in the accumbens and MPOA plummets. Oxytocin is released into the blood and the brain. Endocannabinoids, the brain version of marijuana, make us a little sleepy. Serotonin gushes from serotonergic neurons, inducing a feeling of calm, satiety, satisfaction.

Opioids, such as endorphins, that have been slowly building to heighten appetitive reward now surge, flooding into the limbic system and hypothalamic areas. This is why the heroin addicts Pfaus met in Washington told him that taking it was like sex. As early as 1960, psychiatrist Richard Chessick wrote of "the pharmacogenic orgasm in the drug addict."

(This is where confusion arises over whether or not one can be "addicted" to sex. In contrast to sex-addiction promoters [TV-based celebrity counselors, for instance], Pfaus insists that there is no such thing, and that what looks like sex addiction is really a version of obsessive-compulsive disorder. A man who masturbates five times daily isn't addicted to it; he's become obsessive about achieving five orgasms a day, which is, after all, quite a feat.)

In addition to making us feel good, the opioids of consummatory reward also shut down the dopamine in the prefrontal cortex, allowing executive functions to reboot. As Hartmann put it, "The sky of consciousness gets clear again and gazes in astonishment at the fertilizing rain on the ground." Most people tend to be somewhat less poetic: "My God! Look what we did to the couch!"

Consummatory reward is even more powerful than appetitive reward. "With experience, even as adults, ejaculation and orgasm change the brain," Pfaus explains. "They increase synaptic proliferation in the nucleus accumbens, the cerebellum, and the cool thing is, the accumbens is sensitized to cues associated with sexual reward. . . . Now the neuron changes. You make long-term changes to gene transcription"—the way genes are translated into proteins that bind to receptors to instruct cells—"and you've made a permanent change in the synaptic connection. Now you are Pavlov's dog, and every time somebody rings a bell, you salivate."

When we experience sexual satisfaction and receive the consummatory reward, we (like Everitt's rats that turned on the light because it had become rewarding in its own right) have primed ourselves to receive an appetitive reward from any number of cues associated with that experience. The more often we receive the consummatory reward, the stronger the associations become. What was he wearing? What did she look like? What music was playing? Where was I? These have all become antecedent conditions because the amygdala, which is wired to the accumbens, has logged the circumstances of this very pleasant feeling.

We have developed a fetish. "The mechanisms of diagnosable fetish are just extremes of the very same mechanisms that help people have normal sex," Pfaus says. We are all, to one degree or another, fetishists. Or, put another way, we all, to one degree or another, develop strong preferences.

Reward circuitry and the neurochemicals that act upon it are responsible for why individual people prefer tall, short, blond, dark-haired, skinny, chubby, bespectacled lovers. The sexual reward system has made beauty itself a fetish. How else to explain the roughly $13 billion Americans spend each year on cosmetic surgeries, and the $37 billion they spend annually on beauty products and services? (Worldwide yearly spending is roughly $170 billion.)

This is not only, as some evolutionary psychologists have argued, because human beings evolved to favor the waist-to-hip ratios and facial symmetry that indicate good health and genes favorable for offspring.

"Evolutionary psychologists act like we should never do stupid shit," Pfaus says. "But we do stupid shit all the time! And we do it because we have competing things going on." We wish to breed healthy children, but we also want to satisfy our own experience-shaped reward preferences. "I mean, as far as evolutionary psychologists are concerned the naked body of a reproductive-age woman should be *it*. So why isn't it? Why should we have any preferences? ('Oh, I need you to have hair like a boy,' or 'I need you to wear sexy lingerie,' or 'I need you to talk dirty.')"

Pfaus has illustrated the power of reward-based preference in a dramatic way.

Rats and humans share an innate disgust for the scent of death. It is magnificently repellent, without anyone telling us what it signifies. Rats will do almost anything to get away from the smell, even crossing electrified grids to do so.

Pfaus painted estrous females with a chemical called, appropriately, cadaverine, a synthetic form of the death smell. (If you are ever given the opportunity to sniff cadaverine, pass it up. The stuff can give you nightmares.) He then placed sexually inexperienced males in a cage with these females. The females proceeded to hop and dart like normal. The horrible smell took some getting used to—Pfaus had to train the males on the smelly females many times. But the males did eventually have sex with these females on a regular basis, which goes to show just how motivating a little seduction can be.

Later, Pfaus placed these males with a selection of estrous females, including one he'd perfumed with cadaverine. The males preferred the death-scented over the naturally sweet-smelling (to a rat, that is) females. He even tried scenting some females with lemon, but the males who had their first matings with the deathly females still preferred the nasty odor. Some would mate *only* with females that smelled like death. They had become cadaverine fetishists.

Pfaus's lab has turned rodents into all kinds of fetishists. In the room adjacent to the one in which Parada is stroking rat clitorises, Pfaus walks over to a tray on a counter and says, "So, here's a fetish study." But there's not much to be seen, aside from half a dozen tiny leather jackets. The jackets have little holes, through which a rat can put its front feet, allowing the garment to surround its chest and back. When a rat is trussed up in one, it looks like Marlon Brando in *The Wild One*.

We're sure you can see where this is going. Pfaus gave some male rodents their first experience of ejaculation. "We had two groups, jacket-on, jacket-off"—here, Pfaus stops to laugh, even though he's heard himself and his lab colleagues say "jacket-off" about a hundred times. "For both groups, they are on a female within ten seconds or so, and ejaculate within a few minutes—just like normal." The jacket didn't change their appetitive or consummatory behavior.

But later, when the former Brandos were placed in an arena—sans jacket—with a female in heat, about 30 percent of them refused to copulate at all, and many of the ones that did appear to have sex were really just going through the motions because they didn't intromit (which means they had trouble getting erections). The ones that did copulate took far longer than normal, and the female really had to work for it. "There was failure to activate arousal, simply because the jacket was missing," Pfaus says. "The jacket became the embodiment of arousal. That's phenomenal! We think human fetishists are weird, but whose first experience *isn't* weird?"

Pfaus believes, and we agree, that repeated fantasy and masturbation, perhaps coupled with the general arousal that can come from fear of getting caught or breaking a taboo, seals a fetish in place.

This may seem far-fetched to some, but consider this case history, related by German fin de siècle psychiatrist Richard von Krafft-Ebing:

Mr. V. P., of an old and honourable family, Pole, aged thirty-two, consulted me, in 1890, on account of "unnaturalness" of his vita sexualis. . . . From his fifteenth year he had recognised the difference of the sexes and been capable of sexual excitation. At the age of seventeen he had been seduced by a French governess, but coitus was not permitted; so that intense mutual sexual excitement (mutual masturbation) was all that was possible. In this situation his attention was attracted by her very elegant boots. They made a very deep impression. . . . During this association her shoes became a fetich [sic] for the unfortunate boy. He began to have an interest in ladies' shoes in general, and actually went about trying to catch sight of ladies wearing pretty boots. The shoe-fetichism gained great power over his mind. He had the governess touch his penis with her shoes, and thus ejaculation with great lustful feeling was immediately induced. After separation from the governess he went to paellas [prostitutes], whom he made perform the same manipulation. This was usually sufficient for satisfaction.

Or this, from a 1968 report by Japanese psychiatrists about a twenty-three-year-old man with a vinyl fetish:

Since childhood he had the habit of bed-wetting which, notwithstanding his mother's strict discipline to remedy, continued as late as [when] he left elementary school. In those days, detergents were hardly available, and the patient's mother used [a] diaper to get over the trouble. The patient felt a mixed feeling of shame and pleasure when his mother put him [in] a diaper and cover. . . . As he went to secondary school and began preparing for university entrance examination[s], he once felt a strong attraction to the vinyl raincoat of a foreign woman whom he saw casually in the street. Since then, he began to chase with his eyes after the figures of women clad in [a] vinyl raincoat every time it rained. . . . Later, he could no more be satisfied with only touching in tramcars, and he [bought for himself a] woman's vinyl raincoat, put it on, and masturbate[d]. . . . [These days], he spreads a white vinyl sheet over a mattress and lies down on it [while] wearing a woman's raincoat. Then he further covers himself with a white vinyl sheet as if he [is] put[ting] on a blanket. While he remains quiet, the vinyl softens by his temperature and sends out a peculiar smell, which excites a pleasant feeling in him. Now he fancies himself [in] a masochistic scene in which he himself is a woman and is tampered with in a sexual act by a woman. The sensation of pleasant sexual feeling is small in the absence of vinyl.

Brian has interviewed many fetishists, and they are often able to vividly recall the early circumstances of how their fetish developed. As one tells us, his passion for dressing up in leather and codpieces and behaving as a "master" to his "slave" girlfriend started when he became aroused by comic books. "If I think further back, it's the barbarian slave girl, absolutely. That was one of the first times—that Conan thing—and I was, like, 'That looks pretty good from where I am sitting,' the girl laid up against the man's leg, and the first time I saw that, even at age eight or nine, in the back room of the library, I was, like, 'Rock on!'" And so he did rock on, using the imagery as inspiration to masturbate. (Luckily for him, his girlfriend had followed a similar path, except that she liked the idea of being the slave girl, which proves that there's somebody for everybody.)

Female rope fetishists describe climbing ropes and feeling the rough texture rubbing against their clitorises. Pony-play fetishists talk about riding ponies and feeling sexually aroused. Spanking fetishists recall being spanked as children—which can create general arousal—and placing that spanking into a sexual context. A male shoe fetishist may have masturbated in his mother's closet, surrounded by her shoes, just to pick one possible childhood scenario.

"I would argue that whatever is there at the time of the reward acquires some of the associative strength," Pfaus says.

Pain itself, arousing to the sympathetic nervous system, can become a fetish when it pushes a sexually excited person over the edge, and into orgasm. "Pain can become fixed as part of the sexual act," Pfaus argues. And, as Abraham Alexander found, if Lady Sage, of Columbus, Ohio, is willing to provide some of that pain, you can become extremely determined to visit her, even if you have to embezzle money to do so.

As evidence that it's the reward, not the sex itself, that creates the behavior, Pfaus provided various doses of morphine (an opiate) to male rats, then introduced them to estrous females scented with almond. At higher doses, the males formed a strong partner preference for the females, even though they never had sex because the guys were too high on drugs to mate.

Human fetishists can be driven to destructive extremes, despite their rational understanding that such behavior will harm them. One California man, who had written love poetry to his hydraulic shovel, accidentally killed himself by hanging when he attempted autoerotic asphyxiation from the machine. Alexander certainly knew that embezzling money from the foundation that employed him was wrong, yet he was unable to fight his need to fly to the Buckeye State for discipline. Camp had sentenced many people to jail for drug crimes. Abelard knew the danger he faced. But their rational brains had been shouted down by the power of appetite.

Like Dante, like those who scoffed at Camp's plea to the court, we often ascribe the relative negative or positive outcomes of sex to one's moral fortitude. But one's reaction to the lure of reward can be heavily influenced by how one's brain is constructed and by one's genes. The degree to which rats will learn to assign reward to an otherwise unrewarding act or object, such as Everitt's light, varies according to genetics. To put it in the terms of Pfaus's experiments, some rats are much more likely to become leather-jacket fetishists than others, and those who are most susceptible can become powerfully motivated—so much so that the motivation controls their behavior.

Recent imaging studies in humans have shown that the strength of coupling between the prefrontal cortex and the nucleus accumbens influences how resistant an individual can be to appetitive desires arising from the limbic system. Another imaging study showed that binge eaters, as opposed to ordinary obese people, had a significantly larger spike in brain dopamine levels in response to food cues. The brains of psychopaths can release up to four times more dopamine in response to appetitive cues, such as money, than those of most people. The more dopamine, the greater the drive to achieve whatever the goal may be, no matter the cost—an extreme form of what Loewenstein found with young men in the heat of laptop passion.

Sometimes illness or trauma can throw the craving for reward into hyperdrive. In 2002, doctors in Texas described the case of a man whose multiple sclerosis caused lesions on the right side of his hypothalamus. He developed an insatiable desire to touch women's breasts. A fifty-nine-year-old California man with Parkinson's disease underwent brain surgery and was taking the anti-Parkinson's drug L-dopa, which converts into dopamine. Whether due to the surgery, the drug, or a combination of the two, his doctors reported that "the patient began demanding oral sex, up to twelve to thirteen times a day, from his wife of forty-one years. He forced her to have sex with him despite her serious cardiac condition. He masturbated frequently and propositioned his wife's female friends for sex. . . . He began hiring strippers, and driving around town searching for prostitutes. He spent hours on the Internet looking for sex, and buying pornographic materials. At one point, his wife found him trying to sexually relieve himself while viewing a photograph of his five-year-old granddaughter." Trauma to the prefrontal cortex can also impair that region's ability to say no, leaving the reward system to run wild, just as Judge Camp's defenders suggested.

On the other hand, defects in the reward system can be pathologically inhibiting. Some may find it extremely difficult to act at all because they're immune to reward, obsessively analyzing every possible consequence of a behavior instead. People taking a class of antidepressants called selective serotonin reuptake inhibitors, or SSRIs, can suffer with diminished libido. The drugs keep serotonin available to neurons, which eases the sense of despair common to depression but also quashes sex drive, just as it does in the moments after orgasm.

Most of us are susceptible to reward, and most of us develop strong preferences in response. We'll pay almost any price to obtain our brain's preference, whether that's five dollars for a magazine and its inky eleven-inch reflection of our desire or much more for a visit with a prostitute, drugs for a stripper, or a session with a dominatrix. In all these cases, we're trying to satisfy our craving. And when we find somebody who satisfies that craving, we tend to want to experience consummatory reward over and over. We develop a partner preference as strong as any fetish.

Let's say you have a few orgasms with Bob, and they are very pleasant experiences with smiles and gentle kisses after. From an evolutionary standpoint, you don't need Bob to make a baby; Rodrigo would do just as well. But now your appetite is directed not just to have sex or have an orgasm, but to have an orgasm specifically with Bob, not Rodrigo. You shun Rodrigo in favor of Bob. Bob is your antecedent condition. He's living in your amygdala. You have a partner preference for Bob. You have become a Bob fetishist.

Just as a leather-jacket fetish makes no evolutionary sense, because the true jacket fetishists cannot mate without the jacket, a Bob fetish doesn't make sense, either. Both are "reproductively inefficient." But give the rat jacket fetishists the jacket and they do just fine. Give you Bob, and you do just fine. You're beginning to fall in love. But just beginning. Though appetite and fixation on one person are critical, Larry believes some surprising, gender-specific mechanisms are also required for human love to flower.

4

THE MOMMY CIRCUIT

Maria Marshall has dark mahogany eyes that are simultaneously beautiful and disconcerting. Where one might expect to find seduction or bashfulness or sparkle or curiosity, there are two inscrutable orbs. If you search them, hoping for some sign of engagement, they will dart up and away, wary and uncertain, as if Maria were onstage and had forgotten her next line.

During the first few moments of an encounter, Maria projects the image of a happy, confident young woman. She says "Nice to meet you," and shakes hands. If you ask her how she's feeling or if she's had a good day, she'll answer. But there's an oddly rote nature to her words, and once she has said them, the conversation is over. There's no comeback, no "Did you have a nice drive?" or "How's your day been?"

Maria is twenty-two years old. She lives with her adoptive parents, Ginny and Denny Marshall, in rural eastern Pennsylvania, in a cedar-sided house that sits on three forested acres on the border between Chester and Lancaster counties. Many of the Marshalls' immediate neighbors are Amish farmers who work the land using draft horses and wagons. The setting feels idyllic and safe, one of those places about which people say, "It's a great place to raise children."

Raising children was partly why the Marshalls left Philadelphia for the country and bought their property in the trees. When they discovered that having biological children would require in vitro fertilization, they decided to adopt. And because they wanted what most parents want—to share their love and make a good home—where their children were born, or their race, was irrelevant. So they traveled to Korea to adopt two sons, Michael and Rick. Maria came from Romania, where she'd been born a year before the fall of the dictator Nicolae Ceausescu.

Ceausescu's twenty-four years in power left Romania pocked with the impact of his arbitrary and bizarre diktats. One of his most infamous, handed down in 1966, banned abortion and all forms of contraception. Women were punished for not having enough babies. Romania needed labor, Ceausescu believed, and regardless of his people's personal desires, or even their own poverty, he was determined to raise a bigger crop of workers for the state.

All too predictably, many women abandoned their babies. Children flooded into Romania's orphanages, which were woefully ill equipped, understaffed, and overwhelmed. By the time Ceausescu was shot by firing squad on Christmas Day 1989, his reign ended by a people's revolution, the orphanages had become grim warehouses full of children.

Maria was one of them. She had spent her first twenty-seven months of life in an orphanage by the time the Marshalls arrived in Sighisoara, Romania, the birthplace of Vlad the Impaler, a fifteenth-century prince who inspired Bram Stoker's *Dracula*.

"We stayed in an apartment, and across from our building was an orphanage—not Maria's," Ginny tells us. "From our window, we could get up in the morning and see these two cinder-block buildings, close to each other, and they had windows we could see into. The morning sun would shine through the opposite side of the buildings so we would see all these kids, backlighted, on their knees, rocking back and forth, their heads just going back and forth in the windows."

When the Marshalls found Maria in her orphanage, she was doing the same thing. "The back of Maria's heels were all calloused from hitting her rear end on them," Ginny recalls. "Her bottom would hit the back of her heels. It was smashed down flat. She would do that most of the day."

The children did this in an effort to comfort themselves because they were rarely held or caressed by their caretakers. Some days, the Marshalls could hear children in the orphanage across the way screaming when their heads were shaved haphazardly with old, dull electric clippers—a strategy to cut down on lice infestations and to avoid using shampoo, which was virtually unavailable. Maria underwent the same treatment in her orphanage. Children, including Maria, spent most of the day either on the floor or in cribs that had metal bars, with a flat metal slab or some other kind of lid placed over the top so they couldn't wander out. At twenty-seven months, Maria weighed as much as the typical American eight-month-old.

Photos from the Marshalls' 1990 trip to Romania depict Maria as a nearly bald waif in a tiny dress. In one, a man holds her so the Marshalls can take a picture. Maria's arms are stretched out from her sides like two sticks. Her fingers are spread apart and tensed. Her spine is rigid. There is a look of terror on her face.

"When we held her, she was stiff as a board," Ginny recalls. "She never touched us. It was like holding a plastic doll. She would get stiff the moment you picked her up."

When she says this, Ginny, an outgoing woman with short, light-colored hair and an appealing gap in her front teeth, laughs in a short burst tinged with a smoker's rattle and a note of resignation at the weird absurdity of that image. The Marshalls knew that if they went through with the adoption, they'd be taking on a challenge. But they were determined. "We were the kind of people who said, 'We can beat this,'" Ginny explains. All they had to do was love Maria. What she really needed was a mother and a father. Ginny laughs again, this time at herself.

Maria's remarkably deprived infanthood, almost free of human touch and affection, meant that she didn't know the mother-infant bond. This bond, the first love any of us experience, is the most fundamental one of all. Its evolutionary antecedents are ancient and shared, to one degree or another, among widely diverse species, even some fish. While most fish are content to lay eggs and hope for the best, Amazonian discus fish mothers stay with their offspring, feeding them with a mucus exuded by their skin. The mucus is released at the prompting of prolactin, a hormone that serves much the same function in human women. This interdependent relationship forges a bond between mother and offspring; if you try to separate a mother from her fry she'll thrash and swim in a panic.

But while we appreciate the power of mother love when we see it in people or elephants or fish, we don't often ask *why* a mother cares for her babies. We just assume she does. Yet caring for a newborn is a pretty big change in behavior. It requires an animal or a person to forsake, at least temporarily, its own self-interest in favor of a creature it has never met. Making that change is vital not only for the baby's life itself but also for its future and the future of human societies. We like to think we *choose* to care for our children. And of course we do. But the nature of that choice isn't exactly what most of us expect: as macro as a mother's behavior change can be, it's driven by microchanges in the brain.

Consider sitting in a packed-to-the-gunnels airliner, one row in front of a baby who's crying with the zeal of a La Scala soprano. Most people find this sound annoying at best, dark-thought-inducing at worst. One group, however, is more likely to tolerate the crying, to empathize with the baby, to possibly even enjoy (for a little while, anyway) the child's performance: new mothers. While the rest of the passengers may be fighting urges to strap a parachute to the infant's back and launch it into the atmosphere, new mothers may be tempted to physically soothe and cuddle the baby.

This is because they've undergone a remarkable transformation. "Before I had kids, I never cared much for babies or children," a woman calling herself "SweetnessInFlorida" posted on a dating Web site for single mothers in a typical story of mommy conversion. "I didn't dislike them, I just wasn't interested, and never saw myself as a mother or caretaker of children. I never cooed at babies, never looked at baby clothes and items, never babysat, never went ga-ga over a wide-eyed smiling baby in the market. The thought of diapers made me want to tie my tubes. Then when I had an unplanned pregnancy, and held my baby for the first time, I felt so much maternal instinct and love and warm fuzzy feelings, that I now have two kids and hope for more! I can change a diaper with one hand in the dark!"

Other women feel maternal urges long before they actually have a baby, of course, but sudden converts often marvel at the way babies, once seen as drooling snot factories, become sweet human cupcakes ("I'm gonna eat you up!"). Women who fret about their own diffidence, or outright aversion, toward babies before giving birth can be mystified by the dramatic way their affection for their own baby becomes so all-consuming that they catch themselves regaling their childless friends with sophisticated color analyses of their infant's poo. They describe looking into their offspring's eyes and experiencing a welling up of caring emotion that feels like a tsunami of mother love surging through their bodies.

This reaction to childbirth is a very good thing for the survival of our species, as it is for all mammals. Like SweetnessInFlorida, millions of women every year, perhaps lured into sexual pileups by hormones and brain reward, find themselves "accidentally" pregnant. In fact, roughly one-third of all births in the United States occur as the result of an unplanned pregnancy—as did most of the births on the planet until the advent of modern contraception. Yet, the vast majority of these accidental mothers find themselves happily nurturing a very needy stranger they had no intention of nurturing at all just nine months before.

One could argue, as some do—most notably French historian, writer, and feminist Elisabeth Badinter, who insists the human maternal instinct does not exist—that such nurturing in humans is a choice made under pressure from cultural expectations. It's true that human societies place high demands on mother behavior and that a woman who contravenes those demands will almost certainly face harsh social judgment. Shakespeare, that master observer of human behavior, understood this reflexive reaction to mothers who betray their babies. "I have given suck, and know how tender 'tis to love the babe that milks me," Lady Macbeth declares as she prods her husband to follow through with their plan to kill the king. "I would, while it was smiling in my face, have pluck'd my nipple from his boneless gums, and dash'd the brains out, had I sworn as you have done this." Shakespeare used that declaration, more than all her assassination plotting and husband nagging, to establish Lady Macbeth as one of literature's most loathsome villainesses. And, as you'll see, he even got a lot of the biology correct, right down to the importance of breasts.

WHY MOTHERS MOTHER

Culture may influence whether junior wears a miniature Beau Brummell suit to the first day of kindergarten (or goes to kindergarten at all), and it may have something to say about whether Madeleine will drink a short glass of wine with lunch when she's twelve, but the basics of maternal behavior and the mother-infant bond are inborn phenomena. Mothers are driven to mother by their brains, and the culture of motherhood merely builds itself around nature. As it happens, mammalian pregnancies, fetuses, and then babies are all very good at inducing this mother love by manipulating females' physiological and neural pathways to assure the survival of their newborns.

Rats don't have any of the cultural, social, or religious expectations Badinter and others argue about. Most female rodents don't possess much, or any, active maternal instinct before they become mothers. They don't think they're supposed to "love," or even care for, babies. In fact, virgin rats and mice are usually so afraid of babies they'll either avoid newborns or attack and kill them. Yet even before they give birth to their own pups, they will start making a nest. And once the pups arrive, they care for them, undergoing the same kind of conversion from baby avoider to baby nurturer as many women do.

As early as 1933, scientists noted this transformation in their lab animals. They concluded that there was something about pregnancy and birthing that tipped a female's internal compass, transmogrifying pups from objects of fear to ones of enchantment. Even so, it took a whole generation before biologists began to seriously explore what makes a female turn maternal.

An animal behavioralist named Jay Rosenblatt picked up the mantle by asking the most basic of questions: Why does a mother mother? He placed baby rats in cages with virgin females, and each adult acted according to type: She either kept her distance or made aggressive moves on the little one. She seemed afraid and anxious. Gradually, though, she appeared to lose her fear. After a few days, she'd begin approaching the pup. After roughly a week, she'd perform many of the behaviors typical of a rat mother, like crouching over the pup as if feeding (even though, being a virgin, she made no milk), licking it, and retrieving it if it was removed from her. Clearly, the female rat's brain contained the wiring necessary for her to behave like a mother without actually being one.

In real rat life, of course, pups can't wait a week until Mom decides to start doing her job. A pup, like a human baby, needs lots of care and attention from the instant it's born. Something must activate this existing circuitry before the pups arrive so that mother rats will nurture from the get-go. Taking this reasoning literally, Rosenblatt and Joseph Terkel, then working at Rutgers University, took blood from late-stage pregnant female rats and injected it into virgin rats. The virgins started acting like mothers when introduced to pups.

That was in 1968. Four years later, Rosenblatt and Terkel took the experiment much further. They sewed together the bloodstreams of a pregnant rat and a virgin one, so the factors in the former that would lead to her maternal behavior, even if they shifted over time, would be shared with the virgin. Sure enough, when the pups were born, the mother was attentively maternal, but so was her surgically conjoined twin. And it didn't take the virgin a week to act maternally; she started in right away, effectively providing the pups with two mothers.

Rosenblatt knew hormones had to be involved in this dramatic behavior change, but he didn't know which ones. In subsequent experiments over a period of decades, he, and then his former students, filled in many details of how the mommy circuit works.

During pregnancy, there's a waxing and waning of hormonal tides, much of it directed by cells of the fetal placenta, which essentially hijacks the mother's body to suit its own needs. Progesterone rises and then falls; estrogen rises steadily, peaking near the time of birth. This hormonal coordination has two functions: it prepares a female's body to host the baby (or babies), and it alters her brain.

Late in pregnancy, estrogen stimulates the production of prolactin, the hormone that starts milk production in the breasts, and the receptors for it. It also sparks a striking increase in the number of oxytocin receptors in the uterus. As its derivation from the Greek implies, oxytocin ("quick birth") stimulates the smooth-muscle cells of the uterus to rhythmically contract, forcing out the baby during childbirth. By the time a woman goes into labor, she can have three hundred times more oxytocin receptors on those cells than she did before she got pregnant. Oxytocin is also needed to eject milk from the breasts, so more receptors are made there, too. If all goes according to plan, a rat, or a woman, will be physically prepared to give birth and nurture her offspring at the exact time her babies are ready to come into the world.

But all this prep work in a female's body will be for naught if she isn't motivated to nurture. She has to *want* to mother. Fortunately for baby mammals—and for the young of some other species, like the discus fish—estrogen, prolactin, and oxytocin dramatically change the mother's brain. That change starts early in pregnancy, picking up speed about halfway through, as estrogen and prolactin levels rise.

When a mammalian mother goes into labor, her cervix ripens (a weird term to apply to a body part; it's not a banana). As the vaginocervical area stretches, a nerve signal is sent to the brain's hypothalamus, specifically to the paraventricular nucleus (PVN) and an area called the supraoptic nucleus (SON). (See Figure 1, page 19.) Neurons there fire rhythmically and in unison, signaling the pituitary gland, where pulses of oxytocin are emitted from nerve endings. This oxytocin enters the body, reaching all those new receptors in the smooth muscle of the uterus, and contractions begin. With luck (though we've never heard a human mother describe her labor as easy), the baby is pushed out lickety-split. Today, millions of women every year are given synthetic oxytocin to induce labor.

Prolactin, which has been pumping for at least a few days by now, stimulates the breasts to make milk. But prolactin and oxytocin aren't just released into the body. They act on crucial elements of the mommy circuit, driving maternal behavior in the brain.

As one of Rosenblatt's students, Michael Numan, proved, the hub of this circuit is not a conscious, decision-making part of the brain, but rather the MPOA. He demonstrated this by disrupting it. When he disconnected the MPOA from other elements of the circuit, like the PVN and the SON, mothers would not mother at all. As Numan and others have since discovered, estrogen, prolactin, and other lactogens originated by the fetal placenta physically alter neurons in the MPOA.

When a female rat smells a strange baby, that stimulus becomes information. The information then travels from her olfactory organ to her amygdala, which "reads" overtones of emotion and fear into the novel scents. The amygdala signals other brain regions, which then create reflexive defense or aggression. She'll either back away or lash out at the perceived threat. When pregnancy goes as planned, however, prolactin and oxytocin tip the new mother away from fear. Very near the time of birth, prolactin stimulates the MPOA, which, in turn, signals the amygdala to suppress the fear content of what the mother is smelling or seeing. She becomes calmer, less hyperalert to danger, and more focused on the coming, or newly arrived, pups.

Almost immediately after birth, rodent pups begin to rustle through their mother's fur, looking for nipples. When they find one, they latch on and begin to suckle. Since the nipples contain neurons that project all the way back up to the brain, the sensation of nursing prompts the release of oxytocin into both brain and body. Mother lets down milk. She is calm, focused on the babies, and less likely to be startled by loud noises or other indications of danger.

These peptides are both necessary and powerful in their own right. Without prolactin, rat mothers won't nurture. In 1979, at the University of North Carolina, Cort Pedersen vividly demonstrated that oxytocin could also induce maternal behavior. He injected oxytocin into the brains of thirteen virgin female rats and presented them with a brood of pups. Six of the females instantly became "mothers." They retrieved, crouched over, and licked pups. When Pedersen shot saline into another group of twelve and gave pups to them, none showed maternal behavior. Estrous females responded even more dramatically: estrogen triggers that uptick in oxytocin-receptor density. When Pedersen primed a group of females with estradiol and then gave them oxytocin, eleven out of thirteen displayed the full complement of maternal behaviors.

It's not enough for a mother to simply be willing to nurture. She has to want to, or she won't keep it up. So how does a female, who was so afraid of strange babies just a few weeks before, develop an appetite for caring?

As we've just seen in the previous chapter, the brain has a way of creating appetitive behavior through reward. The MPOA, made very sensitive via estrogen to prolactin and oxytocin, is activated by cues from a female's offspring. Thus activated, the MPOA sends signals to the ventral tegmental area (VTA), where dopamine is made, which dumps the transmitter into the nucleus accumbens. The cry and smell of a pup are now so attractive that a new rat mother will cross an electrified grid to fetch one. Once the rats consummate their appetite and retrieve, lick, or nurse a pup, the reward teaches them what a great idea mothering can be. In rats, it's the initial surge of hormones like estrogen, prolactin, and oxytocin that creates the first burst of maternal behavior. Block those and you can inhibit mothering. When Larry and collaborators in Japan genetically engineered mice with mutated oxytocin receptors, the females became poor mothers. And when Pedersen blocked oxytocin receptors in the VTA, he found he could shut mothering down. But it's the reward that keeps mothers mothering.

Mothering is a social act that involves two individuals. In other words, the mommy circuit is a social one. When Numan disconnected their MPOAs, rats stopped nurturing—but they'd still work for a food reward. So it wasn't reward in general that was short-circuited; it was specifically the reward from interaction with another individual.

When it comes to the basics of the mommy circuit, human mothers aren't as different from rat mothers as one might expect. Just as in rats, estrogen rises, creating the same kinds of hormonal, physical, and brain alterations. Partway through a pregnancy, a woman may begin to feel "maternal" toward her child, even though it hasn't yet been born. She could, for example, become preoccupied with the rituals of making a baby's room, of buying baby gear, of selecting a name, or of obsessively weighing the pros and cons of organic-cotton Onesies.

As a mother's due date approaches, prolactin triggers milk production. During birth, as the cervix ripens, a nerve signal travels up to the brain, and the oxytocin pulses begin.

While rats crouch, human mothers instinctively cradle their newborns in their arms, tucked against their breasts, and babies, somewhat like the rodent pups, naturally begin searching for food. About twenty-five minutes after being born, junior will reach his hand up to his mother's breast and begin massaging the areola and nipple.

As Swedish researchers discovered when they videotaped human newborns' interactions with mothers, babies have a strategy. The scientists monitored the mothers' oxytocin levels and found that this breast massage signals the brain to release pulses of oxytocin—junior is ringing the dinner bell. A few minutes after starting the breast massage, a human baby will stick out its tongue hoping it connects with the nipple. Once it does, the baby will lick the nipple while continuing the massage. The nipple becomes erect. Neurons in the nipple continue to send signals to the brain. Eventually, maybe an hour or ninety minutes after birth, the baby will begin to suck. This also raises oxytocin in both mother and infant.

There is more going on here than birthing and feeding: vital social information is being exchanged.

Let's say you're one of those parents who started an educational savings account and sent away for the Yale student catalog the moment you learned you were going to have a child. Naturally, if you're going to pay Ivy League tuition, you want to make sure you're sending the right kid to New Haven—not the much homelier and duller child delivered by the woman in the next room of the maternity ward. This requires you not only to recognize your baby from among all others, but to give it nurturing priority over the baby next door.

Sheep face a similar problem. As anyone who has walked in a pasture full of them can testify, it can be virtually impossible to tell one from another. This is especially true during lambing season. All lambs look almost exactly alike. In a large flock, with newborns about the same age wandering around, a mother has to know which lamb is the one she birthed. Rodents don't face this problem because their pups are more or less immobile—if they're in the nest, they're probably her babies. In fact, researchers can swap out pups from one litter to another, and the resident mother will adopt them the way "neighborhood moms" tend to feed any kid who drops by. But a ewe doesn't want to suckle just any lamb; she wants to nurture her own.

Ewes manage this mainly with an acute sense of smell, though they also come to recognize their lamb's face. When a ewe gives birth, she regards the placenta of her lamb as milk and honey—which might seem odd for an herbivore like a sheep, but she'll lick her newborn clean, eating much of the placenta and taking in the lamb's aroma.

Human mothers and other primates have a slightly different system based more heavily on vision and hearing. As a human mother holds her baby to her breast, she stares into its face and its eyes, and the baby often looks back. She'll listen for her baby's cries and vocalizations and will vocalize back. She'll constantly touch her baby, groom its hair, and cuddle it. Studies have shown that the higher a woman's oxytocin level, the more she will engage in these behaviors. Whether the input comes from a ewe's olfactory bulb or a human mother's eyes and ears, brain oxytocin helps prime the amygdala with these sensory cues, giving them extremely powerful salience and associating this information with an emotional feeling. This is why a mother's amygdala will be activated in a unique way whenever she encounters her own baby.

Both human and sheep mothers receive brain reward just as mother rats do when they attend their pups. The same dopamine-reward pathway is engaged when a mother senses the sight, smell, and sound of her own baby, linking the sensory cues, the emotional feeling, and the reward, and muting the prefrontal cortex, all of which motivates mothers to nurture. Caring for a baby feels good, especially if it's your baby. Like Jim Pfaus's rats, who overcome their usual aversion to scary odors to have sex, this reward effect leads new human mothers to crave nurturing, and to suspend any former aversions they may have had to saliva, pee, poop, and the presence of an eight-pound stranger attached to their nipple.

Indeed, breast stimulation appears to play a leading role in lending this information such strong salience. An international group that included James Swain, of the University of Michigan, videotaped mothers who had breast-fed and those who had not while they interacted with their babies. They also took images of the mothers' brains as they listened to their own baby cry and a strange baby cry. Breast-feeding mothers, as a group, tended to show a greater activation response in certain brain regions, including the amygdala, when they heard their own baby cry. The brains of non-breast-feeding mothers tended to react to the generic baby's cry in the same way as to their own. This more intense activation of the amygdala in breast-feeding mothers was, in turn, associated with an increase in the affectionate behavior mothers showed toward their own babies when they played together.

The method of delivery also seems to be important for activating this oxytocin-reward axis. Roughly one-third of all births in the United States today are by cesarean section, a procedure that bypasses the birth canal. This bypass prevents nerve signaling from the vaginocervical area from reaching the brain, muting the oxytocin release from the PVN. In an imaging study similar to the breast-feeding test, Swain found that the motivation and reward centers in the brains of mothers who gave birth by cesarean rather than vaginally were less responsive to their own baby's cry. Mothers who gave birth by cesarean also tended to score higher on measures of depression.

Neither of these fMRI studies is proof that a lack of breast-feeding or cesarean delivery is sure to lead to less, or more difficult, attachment between mothers and babies. But they do provide tantalizing evidence that manipulating the breast and stimulating the cervix and birth canal are important for facilitating the mother-infant bond. Interestingly, when ewes give birth to lambs while under an anesthesia that blocks the signal from their cervixes to their brains, oxytocin release is muted and subsequent maternal behavior reduced. Such ewes often reject their own lambs.

Women who look at images of babies while having their brains imaged in an fMRI machine show different activation of their brain-reward regions in response to seeing pictures of their own baby compared with pictures of somebody else's baby. Lane Strathearn, of Baylor College of Medicine, tested mother reactions to happy- and sad-faced babies. When he scanned the mothers' brains, the relative activation strength of dopamine-reward pathways was not only higher when they saw their own baby, but the effect was also especially strong when these mothers viewed their own babies smiling. A happy baby is very rewarding to a mother. The way you keep an infant happy is by holding, touching, feeding, and nurturing it. Of course, no mother wants to believe she's being bribed into caring for her baby, but that's what her brain does.

While the mother is forming an attachment based on information about her baby, the baby is also collecting valuable information—not only about Mother but also about what he or she can expect from the surrounding world based on the way Mother, or another caregiver, is behaving. The baby forms (or doesn't) a bond in reaction to that information. Just as the baby tries to manipulate Mom's nurturing circuits in order to be fed, kept warm, and soothed, it also has the urge to approach. If the system is working properly, the baby finds reward in the warmth of Mother's skin, in the food of Mother's breast, in the soothing sounds of Mother's voice.

Perhaps the most dramatic experiment to show just how powerful the reward of a mother bond can be for a baby was unveiled in 2001 by British scientist Keith Kendrick and his colleagues. They gave newborn lambs to mother goats, and newborn goats to mother ewes. (We'll explain in the next chapter why the ewes adopted the kids.) The babies not only took to their foster mothers, but even if they were raised in a mixed-species group, where they could see and interact with their own kind, their play and grooming behaviors looked like those of their foster mothers: a goat would act like a sheep, and vice versa. Later, when the adoptees were all grown up, foster billies had sex with ewes, foster rams had sex with nannies, nannies had sex with rams, and ewes with billies, which all seems crazy and very unbiblical, but, even after three years of living only with their own species, adopted males still preferred to mate with females of the same species as their foster mothers. After three years, the female adoptees, in a show of organizational flexibility, reverted to mating with members of their own species. In short, the mother-infant bond determined later sexual motivation and action via the reward pathways. Goats became sheep fetishists, and sheep became goat fetishists.

Skeptics of this view of the mommy circuit might insist that humans are different because we aren't so dependent on neurosignaling peptides like oxytocin to push us into nurturing our children. We know we're *supposed* to care for them, that it's the right thing to do. Besides, tests like fMRI experiments are not conclusive, and a lab experiment on humans would be unethical and impossible. True enough. But Maria's experience, and the experiences of other Romanian-orphanage adoptees, is a kind of cruel and unintentional human experiment that not only shows what happens when a child is cut off from mothering but also vividly demonstrates the importance of the signals and circuits we've just outlined.

MUTING THE BOND

"The trees made their tree gestures but human beings were faced by the organized prevention of everything that came natural," Saul Bellow wrote of Ceausescu's Romania in his novel *The Dean's December*. For the first two years of her life, Maria had no opportunity to form any kind of natural bond. And, despite twenty years of loving intervention, including psychiatry, special schools, and enrichment programs, she has exhibited the consequences of that deprivation ever since.

From the moment Maria arrived in Pennsylvania, she was afraid of just about everything. A red aircraft-warning light atop a nearby radio tower terrified her into sleeplessness. Mooing cows elicited frightened squeals. After her hair finally grew out, and it was time for a trip to a stylist for a cut, Maria resisted getting in the chair. Not unlike many other children, she expressed fear during her first barber visit. But, unlike most children, she continued to be afraid, shaking so badly that the stylist finally suggested using scissors to give her a quiet trim. That strategy worked, until another stylist switched on a pair of electric clippers. Maria leaped from the chair, screaming.

At first, Maria seems comfortable and at ease when she and Brian take a short drive. But then, on the return trip, as Brian drives closer to the Marshalls' house, he says, "Let's go a little farther down the road, OK?"

"There's our driveway!" Maria says urgently.

"I know, but let's look at some of the farms."

"You're passing our driveway!"

Half a mile later, Maria says, "I think we better turn around." So Brian pulls into a neighbor's driveway. "I don't think they'll like this," Maria says. As Brian situates the car to head back out into the lane, she says, "I think we should go now."

Among other things, Maria has been diagnosed with obsessive-compulsive disorder. The week before our visit, she packed twenty-two pairs of underwear for a four-day camping trip. "She said she didn't know why," Ginny recalls, "but she worried all week about it." She still has a menu of fears. Electrical cords disturb her. Anything medically related, like hospital beds and stethoscopes, can cause a panic.

At the time of our visit, Maria is working one day a week at Hershey's Farm Market, a local store and bakery, mopping the floor in the kitchen and cleaning cooking tools, mainly. When Brian asks what she'd like to do with her life, she answers, "I never think about the future. I worry for the future, and I get mad about it."

"Your future?"

"Yeah. My future."

"What is it that you worry about?"

"Whether they are going. Whether she [Ginny] is going somewhere, and I worry about that. And what is gonna happen weatherwise."

Because she spends ridiculous amounts of time with anxious rats, Frances Champagne has a good idea why Maria exhibits this kind of extreme anxiety. Rats are naturally anxious creatures, but some of Champagne's rats are much more anxious than others. She first began studying rat anxiety because she needed a job. While a McGill University graduate student in psychology, Champagne worked on a project that attempted to tease out what relationship, if any, existed between pregnancy complications, mothers' stress, and a later diagnosis of schizophrenia in the children. The study ran out of money, so Champagne switched to McGill's renowned neuroscience program because she could earn some cash managing its animal-colony records.

"They were starting to look at how maternal care might lead to long-term changes in development," she recalls. "So I watched rats interact with their babies." She soon realized that individual rats had individual mothering styles. This made her curious about why one rat mother would behave in a certain way around her pups and another rat mother of the same species would behave another way. Champagne's position as record keeper allowed her to trace family lineages. When she sifted through matrilineal lines, she was struck by how the quirks of mothers became the quirks of daughters and the quirks of the daughters became the quirks of the granddaughters. "It was stable, and that still fascinates me—how you can have variations in individuals, and how they transmit it to offspring."

Most of us might suspect the daughters were learning how to behave by watching their mothers, then behaving in the same way toward their own daughters, and so on. That's how we often think of human-family styles. But the truth turned out to be much more disconcerting.

When she was completing her postgraduate work at McGill in the lab of Michael Meaney, Champagne was part of a team led by Darlene Francis (who later worked with Larry) that explored differences in the brains of rat mothers that did lots of licking and grooming (LG) behaviors, and arched their backs around pups. The team found that rat mothers with a larger number of oxytocin receptors in the amygdala licked and groomed much more frequently than rat mothers with low receptor levels. When they became adults, the pups raised by those high-LG mothers also had more oxytocin receptors than their low-LG peers did. The question then became, What caused the variation?

Champagne, a friendly dark-haired Canadian, now works at Columbia University, where her office sits at the end of a hallway in a building fronting Amsterdam Avenue on Manhattan's Upper West Side. The base of her research is located below street level, where scores of rats live in large, often interconnected cages—a rodent version of the city outside. Using such rats, Champagne has traced this inherited variation in behavior not to learning, not to gene mutations, but to how social experience alters the DNA surrounding the genes.

As organisms interact with their environment, genes can be turned down, or switched off, through a process called methylation. When a gene is methylated, a chemical group gloms on to parts of it in the same way a bodyguard runs paparazzi interference by sticking close to a celebrity leaving a Hollywood nightclub. This can make the gene's on-off switch, its promoter, less accessible to the RNA polymerase enzyme that copies the gene's instructions for making a protein. The gene is there, but it's been turned off, or down. The study of this phenomenon is called epigenetics. In addition to a genetic inheritance from our parents, we also receive an epigenetic legacy by the way our parents treat us.

When Champagne took newborn pups from low-LG mothers and gave them to high-LG foster mothers, the pups grew up to be high-LG mothers themselves. Left with their low licking and grooming mothers, low-LG pups grew up to be low-LG mothers. So it isn't exactly genes that make this difference. It's what happens to genes in response to environment. In rodents, this transference of epigenetic inheritance is established for the most part during the first week of life. After that week, females' later behavior toward their own pups is pretty well set: low-LG pups will grow up to be low LG mothers, who make another generation of low-LG pups, and so on.

"For the maternal circuits, it is the rats' experience postnatally, not something they are born with," she says, explaining why some rats become dedicated nurturers and some don't. Then she tells us that, "from the organizational-activational hypothesis, I would say the postnatal period organizes the maternal circuits to be more responsive to estrogen when estrogen levels are high, so that when they become adults, you get activation of those circuits. This is because genes are being methylated to establish a framework for how many or how few estrogen receptors you will have."

Low-LG rat mothers don't do much nurturing because they don't like to. Champagne found reduced estrogen-receptor density in the MPOA of low-LG mothers. Since oxytocin receptors depend on estrogen-receptor activation, those mothers will have lower levels of oxytocin receptors. Lower oxytocin-receptor density results in a more reluctant approach to pups, in lower dopamine release, and in less reward from mothering.

"One of the tightest correlations of neurobiological change, in natural variations of maternal care in the rat, is the nucleus accumbens dopamine output," Champagne says. "That really highly correlates with how much licking they do. Levels go up before they lick, so it's not just the product of their licking that increases dopamine—it is a craving they have. They see the stimuli"—pups— "and it triggers the reward systems."

When Champagne tested high-LG mothers, she found they developed a place preference for the chamber in which they'd interacted with pups— a phenomenon similar to the place preference formed with sexual reward in Jim Pfaus's studies. Low-LG mothers did not form a place preference. If Champagne placed a low-LG mother in a middle chamber between one that held pups and one that held a toy, the low-LG mothers preferred to visit with the toy. In fact, they preferred to be pretty much anywhere the pups weren't, "which is amazing," she tells us. "It's amazing pups of low-LG dames survive at all. The mothers do provide barely adequate care, but if they have a chance, they will get away."

Because the maternal behavior circuits are so integral to other kinds of behaviors, this critical period in a newborn rodent's life has major effects on an animal's personality. Since a pup, or a human baby, gets a sense of the world through its mother, the actions of the mother go a long way toward teaching the newborn what it can expect from the world. To the pup of a low-LG mother, the world is fearsome.

For the first twenty-seven months of her life, Maria was kept alive with barely adequate food and care. But her social experience—the touching, grooming, eye gazing on which a mother-infant bond is built—was almost entirely missing. The world certainly looked like a fearsome place to her.

Pups of high-LG mothers tend to view the world with more confidence. They're more likely to explore and to move assertively in open spaces. If they were people, they'd live in Boulder, Colorado, and go rock climbing.

Conversely, the offspring of low-LG mothers are anxious, aggressive, and stressed-out. "So, if you put them in a novel environment, they will be behaviorally inhibited," Champagne explains. A low-LG rat in an open-field maze will navigate by shakily clinging to the walls. "If you expose them to a stressor," like a loud noise or an elevated walkway, "their stress hormone levels will rise and stay elevated for a much longer period of time" because their oxytocin response, which lowers stress, is attenuated.

As far as the infant Maria knew, the world she entered was a rugged, lonely place. She had every reason to be afraid.

The differences between human and rat reproduction strategies are at the heart of this effect. Rats belong to the Costco school of reproduction—high volume, low prices, low investment. They make lots of babies at a time, and they make litter after litter. People, apes, whales, elephants, sheep belong to the bespoke school. Compared with rodent mass production, we are the Italian tailors of baby making: we take lots of time, and we invest lots of resources. This not only accounts for the existence of the Roddler stroller ($3,500) and certain private Manhattan kindergartens, it also explains why Maria can't seem to truly bond, even with Ginny.

If you are a mouse or a rat, you're every hawk's, snake's, dog's, cat's, coyote's favorite lunch. Fear, anxiety, and hyperalertness come in pretty handy. And because it's so dangerous out there, you want to make as many babies as possible when you grow up. After all, who knows how many of them are going to live? Indeed, Champagne's rats raised in high-stress environments are less maternal because they put all their resources into having babies rather than caring for them. Mothers who engage in less licking and grooming also tend to have high levels of circulating testosterone. Their fetuses are exposed to more testosterone in utero, causing their brains to be masculinized. Champagne believes this primes the female fetus's sexual circuits, at the expense of maternal ones.

High-LG rats do experience stress, and they generate similar levels of stress hormones in response to it. But they cope with it, dialing the anxiety back in a shorter period of time than low-LG rats do. Low-LG pups have a hair-trigger stress response. Once that's activated, those pups continue feeling stressed, which, in turn, exposes their brains to much higher levels of stress hormones called gluccocorticoids.

The same mechanisms work in primates. Monkeys raised in the equivalent of a low-LG environment show the same kinds of behavioral patterns as Champagne's rats. If they're separated from their parents and raised with peers, they're more anxious and respond more quickly and aggressively to perceived threats. Again, this sounds like Maria's instant panicky reaction to anything unpredictable or strange—from the red light on the radio tower when she first arrived in Pennsylvania, to her anxiety over being driven beyond her family's driveway, to what Ginny calls her "anger issues."

THE EMPATHY CIRCUIT

Maria has intellectual difficulties, but she isn't mentally disabled in the ways we'd assume. She's how she is because of how she was treated for twenty-seven months. Studies in the United States and Europe have shown that Maria's conditions aren't unusual for children adopted from Ceausescu-era orphanages. These children—now young adults—exhibit a variety of challenges, not all of which are necessarily caused by a lack of bonding. Malnutrition, for example, may be at least partly responsible for some of the adoptees' intellectual hurdles. But one trait many of them exhibit is most likely directly related to their social deprivation: they have an empathy deficit.

Maria finds it difficult to empathize with anyone, even herself. She often feels sadness, especially alone in bed at night, and she wonders why. "I try to make myself feel happy—a lot," she says. But she can't cast her mind forward to empathize with her future self. This is a skill critical to making goals, to anticipating who and what we'll become, based on how we'll feel in different possible scenarios.

"When you wonder why you are sad, do you ever come up with any answers?" Brian asks.

"No, not really," she replies.

Maria certainly feels emotion. She's capable of limited affection, and can be outgoing and friendly when you speak to her directly. But she often appears to imitate expected reactions rather than feel them.

When Ginny comes down with a headache or a cold, Maria is liable to become angry. "She's not really worried about me getting well," Ginny explains. She is worried about how *she* feels about it. It's about her. It affects her, but not like 'I feel bad because Mom feels bad.' I try to appeal to her rationality to make her understand that it is the appropriate behavior to empathize when somebody does not feel well, not get angry."

"Do you love your mom?" Brian asks.

Maria looks back blankly.

"It's an off and on thing," Ginny interjects helpfully.

"It's an off and on thing," Maria echoes.

It's not that Maria doesn't want to bond with others. She wants this desperately, though she doesn't exactly say so. When she was about twenty, she had a boyfriend we'll call Brad. But Brad broke up with Maria. This made her very sad. She says she has no idea why Brad broke up with her; she can't fathom any reason at all. Ginny knows why, because Brad told her. Maria never seemed to return his affection. She never called him, never invited him to go anyplace, never appeared to be interested in his life. "She could never give anything back," Ginny explains. Interestingly, Brad himself was adopted out of a Romanian orphanage at about the same time Maria was, and he shows many of her traits. He has an especially difficult time looking anyone in the eyes, for example, and when he broke up with Maria, he never looked back. He made a decision, and that was that. Though the two of them still cross paths occasionally, Maria complains that he treats her "like garbage."

After several hours of conversation, Brian asks Maria if she'd like to ask him any questions. "Anything at all," he says. "You can ask me anything about my family, my work, anything, and I promise I'll answer it."

Maria is unable to think up a question.

"Are you curious about me?"

"No."

"Nothing you want to know?"

"Nope. I'm sorry."

Empathy requires some way of detecting other people's emotions as well as a desire to detect them. We use faces, especially eyes, along with language cues to read others' feelings. This skill is first acquired as we cradle in our mother's arms. She looks at our face and listens to the sounds we make. We, in turn, look back. If this behavior, facilitated by oxytocin and motivated by brain reward, doesn't occur, the system for deciphering emotion can either be shut down or become so quiet it can't be heard. (Turning it back on in children with autism is one focus of Larry's research.)

If empathy were viewed as reflecting degree of motivation to nurture, then low-LG rodent mothers could be said to lack it for their pups. But even high-LG rat mothers aren't as attached to their babies as sheep, ape, or human mothers because of their Costco-style reproduction strategy. In both nurturing styles, mothers pay exquisite attention to their babies, absorbing sensory information about them. But rodents, whose most important sensory organ is their noses, do this en masse, taking in olfactory cues about the whole litter and the nest itself, not about individual pups. This is great for scientists like Champagne, who can switch pups from one mother to another without either caring. But it also means that most rodent mothers aren't strongly bonded to their pups, especially in the long term. You could switch out a mother's entire litter, and she'd hardly skip a beat. If you removed the pups, she'd quickly reenter estrus and find another mate with whom she could make another litter. There would be no mourning period, no signs of depression, no pouting.

Human mothers, on the other hand, use the baby-related information encoded in their brains to constantly read their babies' faces and actions to decipher what the baby is feeling. In other words, they develop powerful empathy for their newborns, which, in turn, helps them respond to their needs. But infants raised in the human equivalent of a low-LG situation—a Ceausescu-era orphanage would be an extreme example—don't receive this nurturing.

Studies have proven that oxytocin, and brain sensitivity to oxytocin, enhances the ability to accurately read faces. When people view photos of eye regions on human faces that are expressing some sort of emotion, they correctly decipher the expression more often if they've been given a dose of oxytocin. People who are, for whatever reason, less socially literate at judging how others feel during conversations about emotional events are also better able to empathize if they're given oxytocin beforehand. The silenced oxytocin response in those who were deprived of proper nurturing as babies may result in a blunted ability to calm down when experiencing stress, in less empathy toward others, and in an autistic-like social blindness.

Over a breakfast of blueberry pancakes and sausage at the White Horse, a café in Gap, Pennsylvania, a one-stoplight village in Lancaster County, Maria recalls the time she took up the clarinet. She practiced for a year, but her efforts never amounted to much, and now she can't even play scales.

"Don't feel bad," Brian says. "A couple of years ago, some friends gave me a whole set of blues harmonicas for my birthday, with an amplifier and special microphone, and now, two years later, after lots of practice, I can play a mean version of 'Row, Row, Row Your Boat.'"

Ginny laughs, and so does Maria's brother Michael, but Maria doesn't. Instead, she says, "My gosh!" as if Brian had just announced he'd mastered Tchaikovsky's Violin Concerto in D Major.

Sarcasm is context dependent. It's announced by facial expressions, voice tone, and the disconnect between phrases such as "two years later," "lots of practice," and "mean version"—which all seem to be working up to something grand—and the puny accomplishment of learning a four-note children's song. Maria can't detect sarcasm because she's almost blind to these contextual clues. She has a very difficult time reading the emotion in a person's face, so Brian's smile may have looked like pride rather than self-mockery. She has learned that saying "My gosh!" is the expected response.

Maria's is a literal world. When she was younger, somebody described a group of small children "crying their heads off," forcing Ginny into a strained explanation to a horrified Maria that extreme crying wouldn't actually cause your head to pop off.

The strongest human evidence that Maria's difficulties are based, in part, in the oxytocin and mother-infant-bonding circuit came in 2005, when a University of Wisconsin team studied eighteen children, about four and a half years old, who'd been adopted by American families. Most of the children were born after Ceausescu fell, and during a period when Romania's orphanages were struggling to improve. They'd lived in them for a little over sixteen months on average before being taken to America. Compared with a control group of children, their baseline oxytocin levels were about the same.

Then the team had the children play a computer game while sitting on their adoptive mothers' laps. Every few minutes, each mother-child pair took a break from the game to whisper in each other's ear, to pat each other on the head, and to count each other's fingers—all touching activities. Oxytocin levels in the control-group children rose after they played with their mothers. Oxytocin levels in the adoptees didn't budge.

Of course, not everyone who has difficulty feeling empathy, or who is anxious, spent two years in a Romanian orphanage, and few people show the same kinds of effects as Maria does from her time there. But many people do struggle with a variety of mental and social barriers that might be traced to early nurturing styles, some of which uncannily mirror the variations found in Champagne's rats. For example, when a team that included Champagne's mentor, Michael Meaney, examined the brains of human suicide victims, it found that the victims who'd experienced neglect or abuse in childhood showed the same kinds of epigenetic changes as the low-LG rats. This didn't apply to all suicide victims, just those who were abused or neglected.

Christine Heim, one of Larry's former Emory colleagues, has documented greatly heightened stress responses in women who suffered abuse, neglect, or disturbed relationships early in life with their parents. When she, Larry, and other colleagues tested women with a history of childhood abuse, they found that concentrations of oxytocin in the women's cerebral spinal fluid were lower than in women who didn't suffer abuse. This was especially true if the abuse was emotional rather than physical. Men experience similar effects.

That abuse and neglect would lead to changes in the brain might seem intuitive, but a Dickensian childhood isn't necessarily required to blunt somebody's empathy response. Todd Ahern discovered this when he conducted a particularly provocative study in Larry's lab.

Prairie voles are Costco-style breeders like their rat and mouse cousins. But, unlike rats and mice, prairie vole families are the rodent version of the Cleavers from *Leave It to Beaver*. Most female prairie voles obtain a comparatively high density of oxytocin receptors in some key parts of their brains when they become pregnant and then give birth. As a result, they give their pups lots of attention, with frequent licking and grooming. What's more, Dad doesn't mate with Mom and then quickly skip town to look for another female. He stays, going to work collecting food. When it's Mom's turn to go out to feed, he takes over the home front, cuddling, guarding, licking, and grooming. Grown-up sons and daughters often live at home, even as Mom and Dad are still making new brothers and sisters.

All Ahern had to do to disrupt this social system was remove the father. In fatherless families, the single mothers didn't compensate for their mate's absence by increasing their own nurturing behaviors toward the pups: the pups received less attention. The daughters almost always became pup avoiders as virgins and were less nurturing as mothers. They were less motivated to care for pups. Significantly, both they and the males raised in single-mother families also had difficulty forming adult bonds with mates, making another generation of single-parent families more likely.

Almost thirty years ago, Vice President Dan Quayle raised a very big ruckus by criticizing a TV character's single motherhood. Many social liberals accused Quayle of being ignorant of the reality of single-parent families. Social conservatives insisted he was right. Ahern's vole experiment doesn't decide the issue either way, but what's known is that children of low-affection mothers, or children who've been abused or neglected, as reflected in Meaney's suicide study, show higher rates of depression symptoms and attention-deficit/hyperactivity disorder when they become adults, with girls often becoming low-affection mothers themselves.

This isn't to say that human single mothers are doomed to bring up troubled children. Champagne correctly points out that it may not be an absent father per se that makes for poor-nurturing prairie voles, but the disturbance of the usual family social structure and a lack of compensation for that disturbance. Unlike humans, voles don't have friends, aunts, uncles, grandparents, or day-care providers who can step in and share some of the nurturing load that might have been shouldered by the missing parent. Champagne also argues that even without such social support, human mothers, unlike voles, can use their big cortices to rationalize the need to step up their own nurturing to help compensate for a departed father's time and attention, even when they don't feel like mothering at all.

Even so, there's now a growing body of evidence that the consequences of variation in early human mother-infant bonding do show up later in life, influencing the way we love others, our sexual styles, the way we raise our own children, even the way our cultures and societies develop.

Lane Strathearn first started thinking about how early life affects adult behavior during his pediatric training in Australia. While working on a fellowship studying child abuse and neglect, he became frustrated by looking at damage that was already done. He wanted to know what had led up to the harm, and how to prevent it from happening in the first place.

His first brain-imaging study in the United States appeared to confirm that brain reward was important for mother-infant bonding in humans, as it is in lab animals, but it didn't tell him very much about why some mothers may be less attached to their babies than others. Inspired partly by work from labs like Larry's, Meaney's, and Champagne's, he wanted to know if a human mother's experience of being nurtured as a child influenced her own mothering style.

First, he gave expectant mothers a test called the Adult Attachment Interview, a standard psychology tool. The interview asks people about their earliest memories of childhood, to think of words to describe their relationship with their parents, how one's feelings of anxiety or disappointment as a child were handled by their parents, how parents reacted if one was physically hurt, how they felt when they were separated from parents, and so on. Then the interview asks subjects about the present relationship with their own child and about their hopes for the child's future. The goal is to classify people by attachment styles. Strathearn used three categories: secure, insecure/dismissing, and insecure/preoccupied.

Secure is, well, secure. Strathearn describes an insecure/dismissing person as an outwardly unemotional Mr. Spock–type who is unaffected by other people's displays of feeling. For example, a child raised with admonishments not to cry or be upset might "put on a happy face and soldier on," he explains. A person with an insecure/preoccupied attachment style would do the opposite, acting out emotionally and making decisions based almost solely on the urges of the moment.

After their babies were born, the mothers returned, and Strathearn gauged their baseline blood levels of oxytocin. Just as the orphans in the Wisconsin study showed no difference in baseline levels, there was no difference in them between mothers who tested as secure and mothers who tested as insecure. Then Strathearn had the mother-infant pairs play together for five minutes. Secure mothers showed a much bigger boost in oxytocin than did insecure mothers.

"Only during this period of direct physical interaction with the baby was there this difference in oxytocin response," he says. "That fits together nicely with the rodent models of early-life experiences epigenetically programming the development of the oxytocin system."

When the mothers were imaged in an fMRI machine, secure ones had significantly greater activation of oxytocin-producing regions in the hypothalamus in reaction to pictures of their babies. Key reward-processing centers lit up in all mothers, as they did during Strathearn's first imaging study in which mothers looked at their own babies. But when looking at their own babies' happy faces, mothers with a secure pattern of attachment showed more intense activation than mothers with insecure attachments.

Remarkably, when the mothers looked at their own babies' sad faces, insecure/dismissing women—the Mr. (or, in this case, Ms.) Spocks—showed increased activation of a brain region called the insula. The insula is strongly associated with feelings of being treated unfairly, of disgust, and of physical pain. The reward centers of secure mothers were still active. If this fMRI data reflect what's really going on in the brain, the brains of mothers with secure attachment patterns were telling them to approach their sad, upset babies, while the brains of mothers with insecure attachment patterns were telling them to avoid theirs.

"The feeling I have," Strathearn muses, "is that insecure/dismissing mothers, when they see their babies distressed, instead of their brain signaling an approach response to help the baby, to do whatever baby needs, it activates more of a withdrawal response. It's saying, 'This is painful, uncomfortable, difficult.'"

If these results really do depict an actual cause and effect, this doesn't necessarily mean that those mothers will refuse to help their own children. It means that, on an emotional level, they aren't as motivated to do so. Reward-based learning in these mothers has gone haywire. And if a mother is either dismissing or preoccupied, her baby has to figure out a way to cope with that misfiring system, either by acting out or shutting down.

"I think this is brain organization," Strathearn says. "We talk about 'impairments,' but in my mind, this is adaptive for that infant at that point in time." If the environment is chaotic and disorganized—if, say, a caregiver is using cocaine or there's domestic violence—"in order to attract some attention, the baby needs to ramp up affect, show an exaggerated emotional response, to say, 'Here I am! I have needs!'" Conversely, if a mother or father is cold and dismissing, a baby might try to adapt by becoming cold and dismissing, too.

For babies, this behavior may be adaptive. Later, though, in school, in relationships with lovers, or in relationships with one's own children, it can become maladaptive. "We get the ADHD kid, or the oppositional defiant disorder and other labels we give children, without understanding where the behaviors come from," Strathearn says. "I think, among my colleagues in the medical profession, few have any appreciation of this in their minds. To them, genes are everything. They don't recognize that the environment we find ourselves in changes how genes work."

Such adaptations may affect sexual behavior, too. Maria's extreme experience may have led to an extreme reaction. She reports that kissing Brad—the few times she did so—was OK, but that she hated it when he tried to touch her. She says she "never ever, ugh, no way" wants to have sex. When another young man, whom we'll call Ward, tried to date Maria, he tried to touch her, too. Maria still calls him "Creepy-Hands" Ward.

The opposite outcome may be more common, however. In experiments conducted by Nicole Cameron, also in Meaney's lab at McGill, low-LG rat daughters had more sexual encounters and started having them sooner. As Champagne explains, this response is an adaptation to a stressful environment. "If you're a rat or a mouse, you want to be very anxious and have lots of sex," she says. Such an adaptation may help rat females survive and reproduce. But if they were humans, we'd say they have low self-esteem and daddy issues.

If you recall from the previous chapter, female rodents like to control the pacing of intercourse. Given the chance, they'll choose intermittent stimulation because they enjoy it more and they'll develop preferences for the places and the sexual partners they encounter when they can pace. The female offspring of high-LG rodent mothers demand more time between intromissions. If they don't get it, or if they simply don't want to have sex with a male, they'll act aggressively toward him by standing up on their hind legs like a flyweight and trying to punch him out. In one experiment, high-LG female offspring rejected sex with males half the time.

Daughters of low-LG mothers rarely reject males—only 10 percent of the time, in one experiment. They let the males have their way and won't insist on as much pacing. They'll assume the lordosis position as soon as the male shows up for sex. Daughters of low-LG mothers are easy, and they start having sex earlier than their high-LG peers. Rhesus monkeys raised without their mothers also experience a sharp decline in oxytocin levels, are more aggressive, and are very impulsive, showing little self-control when it comes to seeking pleasurable experiences.

Social scientists have found that human girls from families with disturbed parent-child relationships tend to begin menstruating earlier than their peers. They start having sex earlier, and they aren't as choosy about their partners. As a result, they are at greater risk for teen pregnancy and for eventually passing these tendencies on to their children.

Strathearn suggests that heightened sexual motivation could be another coping mechanism. Those who experienced insecure/preoccupied attachment, he speculates, "may just be going with the flow—whoever comes my way, we'll have sex; whatever happens, happens. There is no thinking of the consequences. For those with dismissing attachment, they may be using sex in a cognitive way, as in, 'I want to achieve this end, and having sex with this person is a means of getting what I want.' There could be the same behavior problem, but different brain mechanisms going on."

He also wonders if family stress and the stress of the modern world in general may be re-creating in humans what lab rodents experience. "It may be a function of the increasing exposure to stress earlier in life that the population is being exposed to," he suggests. "I guess, in humans, stress plays a similar function. If you are in an environment that is stressful and dangerous, the chance of having progeny that survive and thrive is lower, so having kids earlier could be adaptive."

Though Strathearn admits he's guessing, these sorts of corollaries between lab work and human evidence raise fascinating and difficult questions. Champagne, for one, is painfully aware that the results she's getting, and even just the nature of the research itself, are fraught with potential political and social conflict—a prospect that leaves her queasily imagining hordes of protesters marching up Amsterdam Avenue. "I'm glad I work with rats," she declares with a smile. "I can't get in too much trouble with rats." Still, she can't help giving in to speculation about how her work, and that of others, could explain the impact of parenting trends.

While Champagne was still at McGill, she worked on a study of 120 college students. The students reported their perceptions of the maternal care and affection they received as children. Those who said their mothers were low on the affection scale experienced more stress, and for longer periods of time, when they performed a task that induces anxiety.

On the other hand, she says, laughing, some of her lab rats just keep mothering and mothering. "You get mama's boys and not-mama's boys. It's really funny to see the offspring of high-LG mothers, because they just suckle and suckle and suckle, and their mothers don't stop that, and, you know, from their pups' perspective, why bother? If they can get maternal care, why stop?"

These young rats tend to act like nothing bad could ever happen to them, which makes Champagne think of some of her students. Students often seem shocked when they receive a less-than-stellar grade. When they do, Champagne often finds herself fielding a phone call from the parents. "I do get lots of calls. In fact, we professors are not allowed to talk to parents anymore."

"Parenting practices change, and, anecdotally speaking, I can say the students we have now are more dependent and very needy, even in the context of the university, than my generation. I talk with other academics, and they all mention the helicopter parenting going on." (Interestingly, studies have shown that, in some rodents, high LG and oxytocin levels in mothers are associated with their aggression toward others. This may be adaptive behavior, induced by oxytocin and bonding, that prompts some mothers to more vigorously defend their infants.)

She also wonders how cultural differences in parenting and family structure might affect the character of nations, and about the need for government action to help interrupt the cycle of emotional and social deprivation that may be handed down from one generation to the next via molecular changes in the brain. For example, she has shown that early intervention in low-LG rat offspring can change their anxiety behaviors to some degree. An experiment in Larry's lab supports this idea.

In 2011, one of Larry's postdoctoral fellows, Alaine Keebaugh, artificially gave just-weaned prairie voles more oxytocin receptors. When those females grew into adults, they willingly babysat for strange pups, becoming the equivalent of high-LG mothers. Larry surmises that, even in juveniles, the elevated stimulation of oxytocin receptors, perhaps through play with siblings, can continue to shape the social personality.

Strathearn also believes that therapies can help people. "We can help them learn ways to adapt and to compensate," he says. For many, however, it's too late. "The built-in patterns are there, and it is difficult or impossible to reverse them."

One approach that's been shown to reverse them—at least temporarily in limited human tests—involves the administration of oxytocin. Ginny and Denny Marshall tried to help Maria the old-fashioned way. "My husband and I said, 'Well, with love and affection, she will get better,'" Ginny recalls. "And she did, to a point."

Maria doesn't cry often, but that she sheds tears at all is a breakthrough for Ginny. Maria has also managed to fight her own fears. She donated blood, even while feeling overwhelming anxiety. "She had fear like you would not believe," Ginny says. "But she has a fortitude about some things. She wants to do things, even if she is terrified." Maria is also no longer afraid of cameras; she mugs for any available lens. And not long before our visit, during one of what Ginny calls "Maria's meltdowns," Maria actually asked to be hugged. "I was, like, 'Yes!'" Ginny says, laughing a real laugh this time. "I couldn't get there fast enough."

5

BE MY BABY

D r. H. W. Long was tired of the whispering. His medical colleagues had traded anecdotes about patients' experiencing various sexual problems. There were also tracts and books about sex written solely for the profession. But the doctors rarely, if ever, discussed the subject openly with their clients. As a result, Dr. Long believed, the lay public was woefully ignorant about sexuality. Now, though, times were changing. The armistice had just ended World War I—during which the government itself, fearing that so many Americans had been afflicted with venereal disease the army would run short of troops, mounted a very public safe-sex campaign—and old social conventions were crumbling. So, in 1919, Dr. Long published a book for everybody: *Sane Sex Life and Sane Sex Living: Some Things That All Sane People Ought to Know About Sex Nature and Sex Functioning; Its Place in the Economy of Life, Its Proper Training and Righteous Exercise*. It was as plain as any 1960s sexual-revolution guru's guide, including specific and explicit instructions on everything from degree of insertion angles to how a wife should "lift her hips up and down, or sway them from side to side, or swing them in a circling 'round-and-round."

Whoop-de-do!

While acknowledging that there were many possible sexual positions, Dr. Long was adamant about the advantages of face-to-face intercourse.

In this position, face to face (and it should be noted that only in the human family is this position of coitus possible! Among mere animals, the male is always upon the back of the female. They—mere animals—can never look each other in the eye and kiss each other during the act! This is another marked and very significant difference between human beings and all other animals in this regard), it is perfectly natural and easy for the organs to go together, when properly made ready, as here-before described. The woman should also place her heels in the knee-hollows of her lover's legs, and clasp his body with her arms.

The idea, he wrote, was to "excite and still further distend all the organs involved" while husband and wife looked at each other. (We now know that humans aren't really the only ones to have sex facing each other; bonobos do it all the time.)

Joy King rarely has trouble with the "excite" or "distend" part of Dr. Long's advice, but she finds the face-to-face part a troublesome business. King is a former vice president of special projects at Wicked Pictures, one of the world's largest producers of adult entertainment, and now works as a consultant. She started her career in the 1980s, and since then she has become known as something of a savant about what consumers of mainstream porn will buy. It was King who transformed an unknown stripper and wannabe actor named Jenna Marie Massoli into the mainstream-media juggernaut Jenna Jameson.

Wicked Pictures mostly produces so-called couples films. Since this genre eschews fringe and bizarre scenes in favor of tamer, mostly heterosexual, fantasy, King tries to create imagery that will appeal to women as well as to men. She often travels to fan shows, conventions, and retail-store events. She maintains an online social-networking presence so she can talk to consumers, especially women, and ask them what they'd like to see.

King admits it's tough to generalize, but one thing most women agree on is that, while bodies and body parts are welcome, faces are vital. "I met recently with a woman director who is going to be shooting a new line for us, and one of the things we discussed about the marketplace is the importance of eye contact, and shooting eye contact," she tells us. "We need these two people to look deeply into each other's eyes. Oddly enough, she said one of the most difficult things to get her performers to do is look into each other's eyes."

King once tried an experiment of her own by concertedly making eye contact with people she encountered in lines, on the street, in cafés, and through her work. She found that her gaze made most people uncomfortable. "It makes them turn away," she says. Yet when she's been sexually involved with another person, she finds herself doing a lot of eye gazing. Her man will gaze into her eyes, too. It seems not only comfortable but essential. She realized that people don't generally stare into others' eyes—doing so in the animal world is often viewed as a threat—"unless they are engaged in some sort of relationship, and especially people who are in a relationship and having sex."

As unlikely as it may seem, both Joy King's porn problem and Dr. Long's century-old advice directly relate to the reasons why Maria Marshall has difficulty empathizing with others, why mothers look at their babies, and, ultimately, to the genesis of human romantic love.

Like the mother-infant bond, love is a social process. It begins as we begin: with the way our brains are organized, even before we're born. It rides into our brains on the back of lust as steroid hormones activate desire. Tempted by brain reward, our lust becomes more focused before shedding the urgency of raw passion in favor of something deeper, richer, more compelling.

"And wherever she thinks that she will behold the beautiful one, thither in her desire she runs," Plato told Phaedrus of a woman in love. "And when she has seen him, and bathed herself in the waters of beauty, her constraint is loosened, and she is refreshed, and has no more pangs and pains." The resemblance between the philosopher's description of a woman's behavior when she's in love and a new mother's drive to nurture is no accident.

When Ronnie Spector begged her man to "Be My Baby," she had it right. Larry's conviction that love is an emergent property of molecules acting on a defined neural circuit leads to the conclusion that, for women, what we call romantic love is really the result of evolutionary adaptation—a tweaking—of the neural circuits driving maternal bonding, and that our very bodies—specifically, the male penis and the female vagina and breasts—have evolved to help us activate this mommy circuit when we have sex. Arguing that, as far as the female brain is concerned, male lovers are babies leaves us open to sit-com-ready jokes about how men turn into infants when they're sick and hold on to TV remotes like pacifiers. But we believe this explains the remarkable changes in a woman's behavior when she falls in love.

When a young woman in medical school goes to Burning Man, has lusty vacation-sex with a guy who dropped out of high school to run a Web site for vintage Vespa fans, and then announces she's quitting school to travel with him to Phish concerts, she's not fully under the influence of her rational brain. Like the women who sound ambivalent about having a baby, then are filled with baby love, she's been transformed.

True, sex may not be a precondition for what we normally think of as human romantic love. Some people say they fall in "love at first sight," well before having sex. The literature of chivalry is replete with stories of one person's abiding love for another, despite the total absence of physical passion. Today, people form attachments to a coworker, a married friend, or someone who doesn't reciprocate their feelings—people they can never have sexually. Such love-from-afar may exist, but it isn't what we most often think of as burning passion.

Maud Gonne, for example, "loved" the great Irish writer William Butler Yeats. Their relationship lasted nearly fifty years. From their first meeting in 1889 until Yeats's death in 1939, the two of them performed a pas de deux that was alternately romantic and painful.

Gonne—a beauty, an actress, an intellectual, a fervent Irish Republican and Roman Catholic, a believer in mysticism—was never keen on exploring the physical side of love and romance with Yeats. She kept him at bay by insisting that their relationship was one of friendship in the real world, seasoned with spiritual love on the astral plane. Gonne argued that a mystical love was inherently more pure than a base sexual union could ever be. Love at such a remove, she told him, would leave him free to pursue his literary work without the imprisoning sensual and emotional entanglements that can derail one's ambitions.

This attitude left Yeats enormously frustrated. He proposed marriage to Gonne repeatedly, and she refused every time. So he put her, or some avatar of her, into line after line of his poetry. Meanwhile, Gonne rebuffed Yeats's continued overtures toward physically consummating their astral love. All Yeats got was a kiss on the lips in 1899.

Gonne was not so cold with everyone. A brief affair with a French journalist led to the birth of a son who died at age two. After the child's death, Gonne and her lover had sex on his grave. Gonne had hoped that any baby resulting from that union would be the reincarnation of the dead boy. Instead, the mating yielded a daughter, Iseult. (Yeats later proposed to Iseult, too.)

In 1903, Gonne married an Irish Republican, John MacBride. Though the union was unhappy, Gonne didn't turn to Yeats for romance. She continued to regard him as a friend and to address him as such in her letters. In April 1908, she wrote to Yeats from Paris, where she was then living. "My dear Willie," she began. It was the kind of newsy missive one friend would send another. She signed it "Always your friend Maud Gonne."

She wrote another letter in June, to "Friend of mine." When she wrote in July, Yeats was "Willie."

In that July letter, she again wrote of her distress at the thought of a physical union. Compared with a mystical coupling on the astral plane, she declared, sex would be "but a pale shadow." She signed it "Maud Gonne."

When he received this letter, Yeats wrote in a notebook that "the old dread of physical love has awakened in her."

An October letter from Gonne was addressed to "Dear Willie" and, again, signed "Alwys your friend Maud Gonne."

Then, just two months later, in December, Gonne addressed Yeats as "Dearest." She physically longed for him, she wrote. Suddenly, she had such a strong earthly desire for him that she'd prayed to have it taken away. She signed the letter "Yours Maud."

That's quite a change in behavior. After twenty years, how did Yeats go from "friend" and "Willie" to "Dearest"? Why did Gonne find it painful, suddenly, to be away from him?

A few days before Gonne wrote that letter, and after two decades of her insistence on a strictly spiritual love with Yeats, the two had finally consummated their relationship. Apparently, that act of sex emotionally altered her. Gonne now felt for Yeats in a way that twenty years of astral projections had never accomplished.

In 2011, social scientists from several universities, led by Joshua Ackerman, of MIT's Sloan School of Management at the Massachusetts Institute of Technology, explored this kind of transformation. They showed that women are happier hearing the words "I love you" from their male partners after, rather than before, sexual intercourse. The women they studied also said they thought their partners were making the declaration more honestly after having sex for the first time, which implies that the women placed a greater degree of trust in their men after having had sex with them. Being economics minded, the team assigned this change in attitude to (perhaps unconscious) economic reasoning. They hypothesized that once the women had invested an "asset," sex, they were happy to know the investment had paid off. An "I love you" *before* first sex could be a phony sales pitch designed to entice a woman into bed.

This may be true. But there's an unconscious brain process at work that supports the change.

VOLES IN LOVE

In playwright Sarah Ruhl's *Stage Kiss*, a husband catching his wife in bed with another man tells her, "That's not love. That's oxytocin." It's a funny line that reflects the increasingly popular perception that oxytocin is the love hormone, or a love potion. The real story is a little more complicated. We prefer to think of oxytocin as love's doorman.

Much of what the world thinks it knows about oxytocin's "love" effects has come from studying prairie voles. Unless voles are digging in your garden, they're hard not to like. They're rodents, so they are related to rats, but on the rodent-cuteness scale, voles are way up there with chipmunks and baby squirrels. They're small, furry balls—about five inches long, if you stretch them out, with tiny black beads for eyes.

Biologists have long divided themselves into fruit fly, mouse, rat, or worm "people," according to the animal model they use for research. (You'd be amazed by how much some scientists know about fruit fly sex.) There were no vole "people" until around 1980, when University of Illinois animal biologist Lowell Getz hired a young scientist named Sue Carter. Getz had long studied populations of wild voles, mainly because farmers considered them pests. But after Carter arrived, his lab began conducting experiments on captive voles in an effort to figure out their unusual behavior.

When he caught the voles he studied in their Midwest prairie habitat, Getz frequently captured two voles per trap—a male and a female. That was curious enough. Over time, he realized he was often catching the same male-female pairs. Intrigued, he focused more intently on their mating habits, discovering that once male-female pairs mated, they nested and stayed together. This suggested that prairie voles were monogamous.

Prairie vole relationships have strong parallels to human ones. They even "date." When a male finds an attractive female, he courts her. Unlike the perfunctory male rat's style of dating with a few flank strokes to the estrous female, a quick roll in the wood shavings, and a breezy paw wave on his way out to find another female, a male prairie vole is practically Maurice Chevalier. He'll nuzzle and groom his mate, engaging in vole foreplay, and then begin a rather lengthy dance of huddling that can last a day or two. The female plays hard to get. Her sexual behavior circuits force the male to woo her into sex. Unlike other rodents, who come, more or less like clockwork, into estrus every fourth evening, the female prairie vole is a little like the lizards Larry studied in Texas that wouldn't develop eggs until courted. She won't go into estrus until her estrogen is activated by the pheromonal aromas of a courting male. His smell turns her on.

If this were a pair of meadow voles, the male would ejaculate and, without so much as a promise to call later, be on his way in search of another female. Meanwhile, his former lover would wander off to find a nest in which she could give birth and take care of the pups, alone. And when we say "care," we mean barely. Meadow vole females are minimalist moms. They'll nurse, but, after about two weeks, they've had enough, abandoning their pups. They have some serious attachment issues.

Prairie voles, on the other hand, start a real family. The pups use specialized nursing teeth to securely attach themselves to Mom's nipples, and Dad sticks around, cooperating in pup care and protection.

It's not just family formation that's different: prairie voles crave social contact. If they can, they'll spend most of their day with a companion. Meadow voles, on the other hand, are loners, wandering, *High Plains Drifter*–like, from one spot, and one mating encounter, to another.

There are also important behavioral differences within each individual prairie vole, depending on whether or not they've mated. In the wild, virgin prairie voles of both sexes associate freely, spending a little time with individuals of either gender and showing no particular attachment to one partner. They pal around. After sex, though, the new couple will explore for a safe, comfortable home, then set it up to raise their family. The male, despite leaving home to find food, will return, time after time. The mates bond, displaying what Larry believes is the evolutionary antecedent of human love. Their bond is so strong that, in most cases, if the male winds up as a hawk's entrée, his bride will remain single for the rest of her life, rejecting all future suitors.

This emotional and social connection makes them extraordinary among mammals. Monogamy between rodents is extremely rare, and only an estimated 3 to 5 percent of all mammal species live monogamously. Meadow and montane voles are what scientists, sounding as prim as Sister Mary Catherine, call "promiscuous." They mate with multiple partners and don't tend to settle down. Prairie voles are role models of monogamy.

(But they're not exactly the role models some have made them out to be. Contrary to the propaganda of some social and religious conservatives and abstinence-only sex-education advocates, who have often used Larry's name in conjunction with half-baked, misleading, or even outright false pronouncements, "monogamy," for a biologist, doesn't necessarily mean strict sexual exclusivity. Though humans are often considered monogamous, at least by a biologist's definition, some societies, both past and present, don't qualify: Abraham, of the Old Testament; the early Mormons; Muslim males, in some countries; the nineteenth-century Oneida community in the United States; the Toda people of India, in which females are polyandrous; the occasional American religious cult; bohemian enclaves; college campuses; Studio 54 during the late 1970s. In twenty-first-century hookup culture, most humans spend at least some part of their lives not being sexually monogamous. This can be true even if they are emotionally and socially bonded to one person. Certain professional golfers, a raft of politicians, and middle-aged, Lycra thong-wearing swingers come to mind. We'll explore extramonogamous encounters later. For now, for the purposes of understanding how females pair-bond with males, we're talking about an emotional and social connection and not necessarily about sexual exclusivity. It's this bond that prairie voles and people have in common.)

This is a huge difference, not just in behavior but also in social systems between prairie and meadow voles. Yet prairie voles and meadow voles look almost identical and are remarkably similar genetically.

To find out what triggers this pair-bond, Carter set up an experiment in 1994 that proved so startling it became the foundation for the entire field of social attachment. Echoing Cort Pedersen's and Keith Kendrick's exploration of mother-infant nurturing in rats and sheep, Carter injected oxytocin into the brains of female prairie voles that weren't in the mood for sex. Normally, such nonreceptive females would reject any mating overtures and would not form any emotional connection to a male. As expected, when Carter introduced males to these females, the females didn't mate. Even so, they bonded with the males as if they'd had sex. The boost of one chemical in the brain completely altered a vole's life, precipitating the vole version of love and inadvertently kicking off a pop-culture oxytocin craze. (We'll discuss some of the hype later.)

In 1994, under the direction of Thomas Insel (who at this writing is the director of the National Institute of Mental Health), Larry and a colleague, Zuoxin Wang, used voles captured in Illinois to start a colony at Emory University. It has since become perhaps the most famous collection of voles in the world, and there are a lot of them. When Brian first visited the colony, he thought its inhabitants were all exactly alike (a vole is a vole is a vole), but Larry's colony also contains meadow voles. His search for what creates the tremendous behavioral differences between them illuminates why Maria Marshall looks at people with blank eyes, how normal mother-infant bonding is created, and why people fall in love.

Along with Insel, Larry wondered where in the brain the oxytocin might be working to drive the emotional connection in female prairie voles. As we've seen, oxytocin receptors in the brain's reward circuitry help motivate mother-infant bonding. So the Emory team first suspected that prairie voles must have more oxytocin cells and fibers in their brains than meadow voles do. Wang proved they didn't. Insel instead found that the brain areas containing oxytocin receptors were dramatically different in the two species. And Larry discovered that the nucleus accumbens of prairie voles was much more sensitive to oxytocin. The accumbens became a new suspect. The prefrontal cortex, which is connected directly to the accumbens, was also loaded with receptors, so it was a good candidate as well. Larry and the team injected either an oxytocin antagonist (a receptor blocker) or a dummy shot into both these structures, as well as into a control area they figured wasn't involved at all.

Next, they made females receptive by putting them into estrus with estrogen injections, and then arranged a twenty-four-hour date with an experienced male. Afterward, the researchers ran a partner-preference test. They tethered the male to one side of a three-chambered rectangular box, tethered a new male to the opposite side, and dropped the females in the middle. As a group, the females given either dummy shots or anti-oxytocin injections in the control region spent more than twice as much time snuggling up to the original partner. They'd formed strong bonds with their paramour. But the females who'd had their oxytocin receptors blocked in either the accumbens or the prefrontal cortex spent an equal amount of time with each male. They hadn't formed a preference.

This proved that female prairie voles had to have oxytocin receptors in the reward centers in their brains activated if they were to bond with a male. But it didn't prove that mating itself caused that activation.

Later, Larry and Heather Ross, another colleague, created a way to continuously sample oxytocin release into the nucleus accumbens of adult female voles as they interacted with males. They primed the females with estrogen so they'd be receptive. Before matchmaking, Ross took a baseline reading of oxytocin and got, well, nothing. The amount of oxytocin present in the nucleus accumbens was so minuscule that the microdialysis assay they used, which could sniff out anything above .05 picogram in 25 microliters (a picogram is one one-trillionth of a gram), couldn't sense any.

Next, they put a sexually eager male into a mesh cage, then placed this cage inside the female's so the two would-be lovers could interact and do some of the things voles do, like sniff and touch. But the voles couldn't have sex. After two hours, some of the females did show a tiny amount of oxytocin. Statistically speaking, though, there wasn't any real difference between the levels recorded during this restricted interaction and the baseline measures noted at the start of the experiment. The females had no measurable oxytocin before meeting the males, and they had virtually no oxytocin after meeting them in a sexually restricted way.

Finally, Ross released the males so they and the females could do whatever they wanted, including have sex. Many of the males, being males, made brave attempts. Not all of the females were equally receptive. But among those that did permit intercourse, nearly 40 percent produced measurable amounts of oxytocin during mating. When Larry and Ross analyzed the oxytocin measurements from the females that mated and compared them with those that didn't, only the mating female group had significant numbers producing measurable amounts of oxytocin in the nucleus accumbens. Sexual intercourse released oxytocin, which hit the accumbens, part of the brain's reward center involved in the pleasurable feeling derived from nurturing a baby, snorting cocaine, or wearing tiny leather jackets if you happen to be one of Jim Pfaus's fetish rats.

But there was a possible wrinkle: meadow voles must also receive brain reward with sex, yet they don't bond. Wang, now at Florida State University, settled the matter by doing for dopamine release and bonding what Larry and Ross had done for oxytocin release and bonding. Females who had sex raised their brain dopamine output by 50 percent. Of course, just because sex raised dopamine output, it didn't necessarily prove dopamine was required for bonding. So Wang used drugs in different groups of voles that increased the number of activated dopamine receptors, increased dopamine-receptor activation but blocked oxytocin receptors, or did nothing (a placebo). Then he ran a similar experiment to the one Larry had conducted with vole gigolos. It proved that dopamine was required for bonding. Wang even placed anestrous females in a cage with a male for just six hours—which isn't long enough to forge a bond without sex in voles, and these voles didn't have sex—and got strong bonding in the females that had been given the dopamine-receptor activator, just as Carter had gotten with females that didn't have sex but did get extra oxytocin.

In short, both oxytocin and dopamine are required for female voles to bond, and both are released when they have sex. If you recall from the last chapter, both dopamine and oxytocin are also necessary for maternal behavior.

Still, it isn't the presence of these two neurochemicals per se that causes pair-bonding. Prairie voles, meadow voles, rats, and mice all have oxytocin-emitting neurons that originate in the hypothalamus—principally in the paraventricular nucleus. Seemingly disassociated fibers from these neurons drift into other brain structures like loose straw wafting away from a haystack on a windy day. Some of these fibers wind up in the accumbens. The distribution of oxytocin and nerve fibers doesn't differ much among any of these species. So rats and mice and meadow and montane voles that don't form monogamous bonds all have dopamine, oxytocin, and receptors for them inside their heads, and they all also release these chemicals when they have sex. Prairie voles, however, have many *more* receptors for oxytocin in their nucleus accumbens.

But there's one more molecule essential for pair-bond formation. James Burkett, working in Larry's lab, surmised that if bonding involved activating the reward system, it might also require brain heroin—the opioids—just as Pfaus's rats require opioids to form sexual preferences.

As we explained in chapter 3, sex also gives us an opioid shot to the brain; it's this surge that makes sex feel so good. Burkett found that if he blocked opioid receptors, female prairie voles would have sex, all right, but they didn't form a bond with the male afterward. They were missing their hit of brain heroin.

So oxytocin, dopamine, and opioids are all required to initiate the female prairie vole's version of love. Oxytocin facilitates approach. Opioids act on their receptors to create the "wow" of sex. Dopamine helps the brain learn exactly what's causing that wow by imprinting an association between the stimulus—a particular male—and the reward.

Again, mice, rats, voles, and people all have oxytocin, dopamine, and opioids. But there is a big difference between, say, the reaction of a promiscuous female rat and a female prairie vole after getting the dopamine and opioid hits. The female rat associates the wonderful feeling of copulation with the smell of studly males. A male rat associates that feeling with the smell of estrous females. Rats don't care which particular male or female is the source of those smells; any partner that smells right will do. But prairie voles have all those oxytocin receptors being activated in their reward circuits. And as Larry and his colleagues have discovered, oxytocin is more than just a maternal molecule, more than just a molecule that eases the path toward approaching another. It's vital to the ultimate secret of bonding and monogamy. That secret is social memory.

Social memory is different from other kinds, like where we left our car keys or, once we find them, which route to work will bypass the over-turned mattress truck on the interstate. For example, people with an injury to a part of their brain called the right fusiform gyrus (the region that helps us recognize faces) usually have no problem remembering that they hate sauerkraut, that they have a business meeting on Wednesday, or how to advance through all the beginner levels of World of Warcraft. But they can't—if they only see her face—recognize their own mother. Or their significant other. Or anybody else they've met. They have a condition known as prosopagnosia.

Individual humans can thrive, albeit with some awkward moments, with prosopagnosia because society, as well as friends and family, help them compensate for it. The neurologist and writer Oliver Sacks is a good example of this: he has written about his own prosopagnosia and how he copes with it. But a lack of social memory in an animal out on the veld, or in the forest, could be disastrous. If you're a gorilla and you can't recall that the silverback with the mean look on his face is the boss, you could be in for a beating. Flamingos and penguins also depend on social memory. They live in groups of thousands of virtually identical-looking birds. Yet paired mates can pick each other out of the crowd because their brains are good at processing social information and recording that information to memory.

Rodents also need social memory, which they form mainly by smell. When a new animal is introduced to another animal's cage, the resident will engage in vigorous anogenital "investigation." Like a doggie meet 'n' greet, he'll sniff the new guy's butt and genitals. Take the visiting animal out of the cage for ten minutes, then put him back in, and the resident might perform a perfunctory sniff or two, just to make sure he's met this fellow before, but he'll recognize him and then go on about his normal rodent business.

The thing about rats and mice, though, is that while they do have social memory, unless something extraordinarily shocking happens—a vicious fight, say—it's pretty lousy. Like glib Hollywood agents, all kissy face and best pal with you during the first few minutes of champagne-glass clinking at the Golden Globes, who then forget who you are when you call the next day to pitch that great screenplay idea, rats and mice quickly forget. Keep the new animal out of the cage for an hour instead of ten minutes, and the resident male will vigorously investigate again because he now has no idea who this animal is. Female rodents have better social memory than males, but they eventually forget, too.

Larry's colleague Jennifer Ferguson proved that it's oxytocin, acting in the amygdala, that plays a critical role in the ability to form memories of individuals. First she gave mice the equivalent of prosopagnosia by knocking out their oxytocin gene through genetic manipulation. She got mice that whizzed around mazes they'd learned and found food in places they'd found it before but that could never remember another mouse they'd just encountered. They had zero social memory. If the mouse was perfumed with lemon or almond scent, the mouse equivalent of a name tag, they did better, but that's because they were recalling the unnatural scents, not mouse social cues. It's like Oliver Sacks's failing to remember an old friend at a convention, but then glancing at his name tag and heartily shaking his hand.

When Ferguson injected oxytocin into the genetically altered mice, their social memory was restored. Giving oxytocin to the mice *after* they'd been exposed to another mouse didn't help at all. Just as Michael Numan, whom you read about in chapter 4, found, when he disconnected rat MPOAs and mothers stopped nurturing but would still seek a food reward, oxytocin was critical to processing social information, but not other kinds.

After a quarter century of investigation, it's now clear that vole "love" and monogamy come down to a defined set of ingredients: a feel-good reward and a strong, emotionally salient memory of the individual one was affiliating with when the reward hit. That's the cocktail.

CROSSING THE BRIDGE TO PEOPLE

Many people, including scientists, once thought (and many still do) that such research on furry little creatures was fascinating—but certainly not very applicable to humans. Our big, rational brains had managed to divorce themselves from animal drives, they argued. Those who believed this relied on one undisputed fact: there weren't any tests in humans. And, if you couldn't prove that love in a vole was somehow parallel to love in people, the discussion was stalled.

Markus Heinrichs didn't set out to address love, exactly, but he did want to bridge the gap between animal experiments and human beings. When Heinrichs, then a young doctoral student at Germany's University of Trier, and now head of the behavioral psychology department at the University of Freiburg, read the animal studies, he began agitating to give oxytocin to people to see if it had any of the same behavioral effects in them.

"But nobody wanted to do it," Heinrichs recalls as he sits in his new Freiburg office. "I went to my supervisor and said, 'I would like to test this,' and he said, 'No, nobody does that.'"

There were theoretical reasons why nobody did it, but there was also a powerful practical one. Since hormones like oxytocin have effects in both brain and body, injecting them into the bloodstream could trigger unwanted side effects in the body—and because the molecules are big, they don't appear to pass through the blood-brain barrier, so it'd be pointless to inject them anyway. Besides, few thought one molecule could have a strong behavioral effect on people.

Heinrichs, a tall, friendly, boyish-faced man who, perhaps to lend himself some gravitas, favors the old-timey round spectacles of an early twentieth-century Mitteleuropa thinker, was convinced that oxytocin could help people with mental disorders like social anxiety or autism—or at least elucidate some of the mechanisms behind those puzzling conditions. Determined, he spent five years going to science conferences and nagging scientists in the field, including his supervisor, until he made such a nuisance of himself that he wore down the opposition. One expert said, "If it made any sense, someone would have done it in the United States!" he recalls, throwing back his head with a guffaw. Finally, he was allowed to conduct a small trial.

Since there was no way to ensure that the oxytocin given to a human would enter their brain, and out of fear of possible harm, the trial used lactating women because breast-feeding was known to naturally increase oxytocin levels. Women were assigned to either nurse or simply hold their babies before they underwent a stress test. The test, created at Trier, requires subjects to speak about themselves before a panel of stone-faced interviewers, and then to count down from a high number in increments of seventeen as the interviewers abruptly correct any mistakes, and urge subjects to speed things up. (Try it; it's pretty stressful.)

The women who breast-fed showed a significantly lower level of anxiety under stress. Heinrichs concluded that, in the short term, at least, the act of suckling suppressed the usual stress response. Since the levels of oxytocin in the blood didn't change in either group of women, the stress-response difference had to be accounted for by increased brain oxytocin brought on by the breast-feeding. Exactly the same thing happens in lactating rats and voles. Unless there's a serious danger, nursing rats will ignore stimuli that would stress them at other times. Nonlactating rats will also have a blunted stress reaction if they're given a shot of oxytocin into their brains.

In 1996, German scientists reported a way around the blood-brain barrier, but their method didn't really get noticed until 2002, when other Germans published the same solution. It was an almost comically simple technology. They formulated nasal sprays containing peptides and found that after spritzing them up the nose, the drugs would appear in cerebrospinal fluid—the conduit from the brain to the body—about thirty minutes later.

By then, Heinrichs had moved to the University of Zurich. But in a continuation of his work at Trier, he gave men either sprays of oxytocin or a placebo and then conducted the Trier Stress Test. The results echoed the study on the lactating women.

Although he'd struggled for a long time to conduct human oxytocin experiments because he thought results like those Larry had with animals could illuminate a path toward helping people, Heinrichs could hardly believe his own results. "We got such strong effects that we worried they would not be reproduced. They were very strong behavioral effects."

He needn't have been concerned. His first results with intranasal oxytocin have been reproduced and expanded upon over and over again. Science has shown that the animal findings have strong parallels in humans.

Social memory was one of the first ingredients of vole love to be explored in humans through the use of intranasal oxytocin. While rodents rely mostly on scent to distinguish a familiar individual from a stranger, we depend more heavily on our eyes. We use our eyes in combination with social memory to decide if the person we're seeing is a friend from work, our husband, or our mother, and we also use these tools to divine other people's moods and intentions. Is he flirting? Is she angry? Does that guy want to mug me? Is my baby in distress?

When psychologist Adam Guastella, of the University of New South Wales, asked two groups of volunteers to look at pictures of human faces that displayed a neutral expression—faces that weren't particularly angry, happy, or sad—he found that the group given an intranasal dose of oxytocin spent far more time fixating on and gazing at the eyes of the human faces, even if they were instructed to look only at the mouths. Oxytocin apparently biased their brains to direct their eye gaze to the eyes of others. Absent any obvious signals of emotion or intent on the faces, subjects were trying to mind read—to infer the emotions of others—by staring into the windows of their souls. The volunteers given a placebo spray did not fixate so much on the eyes.

A Swiss experiment tested the ability of men not only to recall a face they'd seen before but also to *know* that face. The men were divided into two groups. One group received a spritz of oxytocin, and the other group received a placebo spray. Then the scientists showed the men a series of images that included houses, sculptures, landscapes, and faces. Some of the faces looked negative, some positive, and some neutral. The men were asked to rate the approachability of each object or face, a task designed to make sure they got a good look at each image.

The next day, the scientists surprised the men by showing them the same series of images, but with new faces and new objects thrown in to create visual noise. Then they asked the men to tell whether or not they recalled the image exactly, recognized it vaguely but weren't sure of its context, or regarded the image as new. When it came to recalling the objects, there was no difference between the men who received oxytocin and those who did not. But the oxytocin group was better at remembering faces—the flip side of the oxytocin-deficient mice that remembered all sorts of information but not other mice. What's more, there was a dramatic difference in how the two groups of men rated the faces' familiarity. The placebo men often said they recognized new faces when, of course, they hadn't seen them before. The oxytocin men didn't make this mistake nearly as often: they were better at discrimination.

People given oxytocin are also more likely to communicate with intimates in an open and positive manner. Take "Couple 35," for example. The fellow—about thirty, wearing black pants, a watch, a black-and-white shirt—has suddenly adopted the universal guy hangdog look (slouching, slumped shoulders, bowed head) that says he'd rather be getting a tooth drilled than sitting in a room having this conversation. His longtime girlfriend, also about thirty, in jeans and a green sweater, hasn't said anything to elicit this reaction. It was the way she said it. When he claims she was too "controlling" when all he wanted to do was go out with his friends, she replies that she'd be more than happy to go with him if he would just do something that interested her once in a while, which might seem to be an accommodating thing for her to say, but we, like the boyfriend, are noticing her chin jutting out, that she's rolling her head, and using a slightly mocking tone of voice. What's more, he is extrasensitive to all her cues because he has just received a spritz of oxytocin up his nose.

We're sitting in the dark, in a small University of Zurich lecture room. And though we're watching a video, we're feeling a little sorry for the hangdog young man in it—until we remember that he's asked for it. He volunteered. Beate Ditzen, a professor of psychology, a former student of Heinrichs's, and one of Larry's former collaborators, put this couple in the room and asked them to discuss a topic of relationship conflict for ten minutes, so the fellow knew what he was in for. Ditzen wanted to find out if oxytocin actually has any impact on real-life human relationships—not just on lab economics or eye-gazing-at-pictures tests, because, in addition to her academic work, she counsels couples.

After a few minutes, Ditzen stops Couple 35 and skips to Couple 31. (She tested forty-seven heterosexual couples in all.) They've received a placebo spray. For ten minutes, they talk about that couple-conflict evergreen: housework. The woman claims that he doesn't like it, so he doesn't do it, leaving it to her. He replies that he'd be happy to do it, but . . . And then there's some flimsy male dissembling about schedules and lack of skill in the technically complex art of dish washing.

This couple's body language and tone of voice look almost exactly like the social cues of Couple 35. We can't tell who got the drug and who got the placebo. Neither can Ditzen; she has to look that up later in her records. Given oxytocin's popular image as a love potion, you'd expect the drugged couple to fall into each other's arms and kiss their way out of conflict, but the effects of extra oxytocin are much more subtle. It was only after an independent third party carefully analyzed the videos for "positive" behaviors—eye gazing, smiling, touching, praise, openness— and "negative" ones—withdrawal, criticism—that Ditzen got very clear results. Individuals within couples who received the oxytocin exhibited more positive than negative behaviors toward the other. Among the couples that received the placebo, individuals displayed more negative than positive behaviors. Again, oxytocin biased the brains of human beings toward the positive in actual relationships. It also, as in previous studies, eased stress, greatly lowering levels of the stress hormone cortisol. This meant that the oxytocin couples were less guarded and more open.

As if to confirm Ditzen's results, Adam Smith and Jeffrey French, at the University of Nebraska, tested monogamous marmoset couples. When Smith squirted mists of oxytocin up the noses of some marmosets and not others, the ones receiving the squirts spent more time huddling with their partners. But when marmosets were given an oxytocin blocker, they wouldn't even share food with their mates.

Given results like that, it's not surprising that Israeli scientists in early 2012 reported that oxytocin levels could predict relationship success. They measured oxytocin in new couples, then followed those couples for six months. The lovers who were still in the original relationships after six months had the highest levels at the start.

Oxytocin not only nudges the brain toward positive social communication and encourages mind reading but also actually improves the precision of our internal emotion detectors. "I am always astonished, still, that, with one single dose of oxytocin, you will be 50 percent better at reading emotions," Heinrichs marvels.

While accuracy of reading emotions improves with oxytocin, there does appear to be a bias. When the brains of people given oxytocin were examined with fMRI, amygdala activation differed, depending on whether the people were looking at fearful or happy faces. Activation was somewhat blunted in response to fearful faces and enhanced in response to happy ones, suggesting that oxytocin tips the brain toward paying attention to positive social cues like smiles and twinkling eyes, and away from negative cues.

Other experiments have shown that both men and women given oxytocin rate male and female faces as more trustworthy than do people given a placebo spray. In addition to faces seeming more trustworthy, male faces appear to be more attractive to women dosed with oxytocin. After Heinrichs published his first study on the effects of an intranasal dose, some colleagues in Zurich's economics department ran a test using real money exchanged between "investors" and "trustees" (like an investment broker). They found that subjects given the real spray handed over more money to trustees than did those receiving a placebo. They trusted more, akin to the women who were more open to hearing "I love you" from men after having had sex with them for the first time. The effect disappeared when subjects were asked to invest with a computer. Oxytocin does not universally lower barriers to risk taking. It only works in person-to-person situations.

Subjects in all these tests had no idea what was sprayed up their noses.

As the subtle differences among Ditzen's couples suggest, they're not conscious of any behavioral effects. "We always ask all the people if they had placebo or oxytocin, and they never get it right," Heinrichs reports. "Ten years ago, I would have said I doubted that one hormone was doing all these things. Now there is too much evidence."

This heightened awareness of social cues, along with a positive bias, tends to ease people into relationships with others by lowering apprehension—just as it eases new mothers into approaching and nurturing their babies and drives voles to affiliate with each other, leaving them vulnerable to a hit of sexual brain reward.

The effects, however, are not uniform. Maria Marshall's experience is a drastic example of how human variation in the oxytocin system can play out. And there is mounting evidence that slight differences in the gene containing the recipe for oxytocin receptors can help explain more typical variations in love and bonding among individual people.

Israeli scientists asked men and women to participate in an economics exercise called the "dictator game." It's not really much of a game. One person, the dictator, is given a sum of money (in this case, 50 points, with each point equaling one shekel) and told to decide how to split it with a second person. The second person has to accept the decision. That's it. But the game has been shown to be a pretty good indicator of altruism— the more money shared, the more altruistic the dictator is. The scientists also used a second exercise, called "social value orientations," that tests one's concern for others. As you might expect, the more concern subjects showed for others, the more they shared. Both those measures were significantly correlated with a particular version (single-nucleotide polymorphism, or SNiP) of the oxytocin-receptor gene. Those who didn't have that version showed less concern and shared less money. When the scientists tested only women, they got even more robust results. Women with one version of the gene gave an average of 18.3 shekels to the other person. Women with the other version gave an average of 25.

Would such natural variation in one gene make any difference in real-life love? Using a large database of male-male and female-female twin pairs, Swedish researchers made the connection. Women carrying one particular SNiP were more likely to be less affectionate with their spouses, to have a significantly higher chance of experiencing a marital crisis, to have lived through more social problems as young girls (echoing the effects of low-LG nurturing in rodents), and to have somewhat worse communication scores on an autismlike behavior measure. Men seemed unaffected.

"Interestingly," the scientists wrote, "this is in line with vole studies showing that non-monogamous vole species show less affiliative behaviors than monogamous vole species already in their first days of life."

WHY MEN HAVE BIG PENISES, WOMEN HAVE BIG BREASTS, AND A WOMAN IS LIKE A SHEEP

Men and women don't spray oxytocin up their noses in order to approach and then trust each other, enter a relationship, and fall in love. (Though that hasn't stopped some unscrupulous marketers from trying to sell "oxytocin" to barflies.) Instead, we need mechanisms that will spike our brain oxytocin the way it does in new mothers. Lucky for us, we have them. So do sheep.

Lambing is a critical time for sheep farmers. Whether a year will be profitable or not depends on how many of the lambs survive. When lambs die, they typically do so in the first three days after birth—often because their mother has died or, for some reason, rejected and refused to nurse them. In either case, it would be a boon to sheep farmers if they could figure out a way to make ewes adopt orphaned or rejected lambs. But, as we've explained, ewes don't want to nurse any other lamb but their own. And they can tell their lamb from all others, even in a field crowded with newborns, by the strong sensory and emotional imprint—their social memory—that their baby has made on their brain from all that sniffing and placenta munching.

Once upon a very long time, a shepherd figured out how to turn ewes into adoptive mothers. Exactly who this shepherd was, and how he discovered this trick, has disappeared into the mists. Perhaps it's better for the shepherd's reputation that we don't know his identity, because his trick involved stimulating the vagina and cervix of a ewe that had recently given birth. Exactly how he stimulated the vagina and cervix of the sheep is a mystery, but, presumably, he noticed that when he did so, with an orphaned lamb nearby, the ewe would allow the strange lamb to nurse.

Keith Kendrick and Barry Keverne, his University of Cambridge colleague, pioneered a lot of the science behind sheep bonding. It was Keverne who figured out why this old shepherd's trick worked.

In 1983, Keverne and coworkers went shopping for unusual lab supplies. They entered a sex shop and soon walked out with a very large dildo, which goes to show that TV's Mr. Wizard was right: You can use all kinds of household objects to do science. Anyway, they used the dildo to stimulate the vagina and cervix of a ewe and found that even a ewe that had not given birth (but that had been primed with estrogen and progesterone to mimic pregnancy) would exhibit "the full complement of maternal behavior . . . after five minutes of vaginal-cervical stimulation" (VCS). New mothers didn't require any hormonal priming. Kendrick found that five minutes of manual VCS could make a ewe bond with a strange lamb— even after it had already bonded with its own—for up to twenty-seven hours after it had given birth. A lot of scientists proceeded to stick their hands and "lab tools" up lots of other animals' vaginas, to find that VCS works in goats and horses, too.

In other words, Kendrick and Keverne discovered that mimicking the natural stimulation of giving birth triggered ewes to bond with a strange lamb, whether or not they'd bonded with their own and whether or not they had actually given birth at all. Birth could now be dissociated from mother-infant bonding: it's the stimulation of the vagina and cervix, and the release of oxytocin such stimulation causes, that touches off the bonding process.

In fact, though it doesn't make nearly the good story that VCS does, you can induce bonding in a ewe simply by giving her a shot of oxytocin to the brain, specifically to the paraventricular nucleus (PVN), the area where human (and sheep) brain oxytocin originates. Give oxytocin to a hormone-primed ewe, and she will adopt an alien lamb in less than a minute.

VCS will also supercharge a female rat's social memory. If you give a VCS massage (in case you're wondering—and why wouldn't you?—experimenters use a probe, not their hands, on rats) to a female either about to go into or already in heat, then introduce a juvenile, the adult female will still recognize that youngster five hours later. This proves that stimulating the vagina and cervix creates oxytocin release in the brain, inducing reward learning and enhancing social memory and the desire to bond.

Nerves link the genitals to the oxytocin-producing PVN of rodents, a fact discovered by Larry's wife, Anne Murphy, and her coworkers. The influences of these signals on oxytocin-producing cells is greater in females than in males, which makes sense, since estrogen increases the expression of oxytocin receptors. Murphy believes it's likely that oxytocin-producing cells project into parts of the brain regulating social behavior, in effect tying rodent sex and rodent love together.

Experiments like the ones Larry and his collaborators and colleagues have performed that demonstrate voles having sex naturally get hits of oxytocin, dopamine, and opiodes in their brains—and that these chemicals drive vole love—are impossible to perform in humans. So we can't prove that Maud Gonne changed her attitude toward Yeats because the two had intercourse. But there's little doubt that oxytocin is released in the human brain during sex and orgasm.

It's known that oxytocin is emitted into the blood when people have sex. For centuries, midwives have told women that labor can be hurried along by having intercourse. In modern times, obstetricians have used dilators and water-filled balloon devices to accomplish the same job. The reason why these devices work and why intercourse can initiate labor is that both stimulate the vagina and cervix, just as in voles having sex and ewes undergoing VCS. It's also highly probable that oxytocin is released into the brain at the same time because, as Murphy showed in rodents, the brain's oxytocin-producing center is directly connected to the genitals.

Humans appear to have evolved specifically to exploit this circuitry.

It may come as welcome news to many men that, relative to our body size, we are very well endowed. We have the biggest penises of any primate. The average erect gorilla penis measures only about an inch and a half. (Don't get smug. Barnacles have penises up to eight times the length of their bodies.) It's also true that size does matter.

To twist the vertically gifted Abraham Lincoln's answer to a man who asked how long a person's legs should be ("Long enough to reach the ground," Lincoln said), for reproductive purposes, a penis needs to be only long enough to make a sperm deposit near the opening of the cervix. The human vagina has a mean depth of about 63 millimeters (or about 2.5 inches), as measured from the cervix at the back to the introitus, the spot where a hymen would be, at the front. Given that the average non-porn-star's erect penis is about 13 centimeters long, or 5 inches, the human penis would seem to be a case of evolution overdesigning our equipment. The vagina is incredibly flexible, which is how an eight-pound baby can travel through it, so it will accommodate even the freakishly endowed, especially when a woman is sexually excited. But the fact remains that most men have penises that are significantly longer than necessary to put sperm in the right spot.

Evolutionary theorists have long questioned why we have longer penises than our primate cousins. One theory has it that men use them for a kind of Anthony Weiner–esque "look-at-me!" display—sort of like a lion's mane: a symbol to other males that we are he-men to be reckoned with. Our ancestors with longer ones chased rival males away from females. Another theory suggests that longer penises evolved in men because our sperm compete with that of other males, who may mate with the same female just after us, inside the vaginal canal. The closer to the cervix we can make a deposit, the bigger head start we can give our swimmers in the great race to the egg. A third idea suggests that women are choosy. Unlike most female primates, women can orgasm more than once. When they figure out that the male penis is a handy device for inducing said orgasms, they select men who are generously equipped.

We believe there's a better way to account for big human penises; namely, that Dr. Long knew what he was talking about way back in 1919. Larry believes the human penis has evolved as a tool to stimulate the vagina and cervix so oxytocin is released into a woman's brain. The bigger the penis, the more effective it is at triggering an oxytocin surge during intercourse. Surges of oxytocin help ease any apprehension or anxiety a woman may have, making her open to her lover's emotional and social cues. She takes in his face and eyes and powerfully registers the emotional context in her amygdala. Presumably, dopamine and opioids are released. While she stares at her lover in a way that would be disconcerting in other contexts, she receives pleasure, associating it specifically with him the way a mother keys on her baby. This is a far more erotic and pleasing scenario than a shepherd manually stimulating a ewe to adopt a lamb, but the mechanism is about the same.

A study conducted by Stuart Brody, of the University of the West of Scotland, has indicated that, as fun as they are, neither oral sex, nor masturbation, nor any other form of sexual activity gives women the feeling of overall relationship satisfaction, including "feeling close to one's partner," that penile-vaginal sex creates.

We realize we may sound like advocates for the missionary position, but face-to-face sex also has another advantage: a woman's breasts are very handy.

In chapter 3 we argued that breast obsession was innate in the heterosexual human male and that Hugh Hefner got rich exploiting it. From our earliest days, breasts have been central to human erotic imagery.

Men don't just want to look at them, though. We clearly love to play with them. We lick them, nibble them, twist them like radio dials. We sing "My Way" into them as if they were microphones. We don't have to be having sex to do any of this, but we do it especially during sex, and we're the only male animals that do. This makes no reproductive sense.

Yet breast play during sex is nearly universal. When Roy Levin, of the University of Sheffield, and Cindy Meston, of the University of Texas, polled 301 people (153 were women) about breasts and sex, they found that stimulating breasts or nipples enhanced sexual arousal in about 82 percent of the women. Nearly 60 percent explicitly asked to have their nipples touched.

Like male penis length, this human boob eccentricity has long puzzled evolutionary biologists. Some hypothesize that full breasts store needed fat, which, in turn, signals to a man that the woman is in good health and therefore a top-notch prospect to bear and raise his children. But men aren't known for being so choosy about sex partners. If the main goal of having sex is to pass along one's genes, it would make more sense to have sex with as many women as possible, regardless of whether or not they look like last month's Playmate.

Another hypothesis is based on the fact that most primates have sex with the male entering from behind, just as Dr. Long stated. This may explain why some female monkeys display elaborate rear-end advertising. Men, goes the argument, needed a little erotic enticement to remind them of how our evolutionary ancestors used to do it. So the two breasts in front became larger to mimic the contours of a woman's rear.

But there's only one neurological explanation for this oddity among humans. As we saw in the previous chapter, newborns engage in fairly elaborate manipulations of their mothers' breasts not only to bring down milk but also to release the oxytocin that facilitates bonding by encouraging the mother to emotionally approach her baby and imprint its sensory cues into her amygdala. The baby also gets brain reward and the feeling of security. Male breast fascination begins there.

Later, in a recapitulation of those earliest days, we use breasts to help create and maintain the romantic bond. Breasts, like penises, have evolved into tools for stimulating oxytocin release via the mother-infant bonding neurocircuit.

Manipulating the breasts releases oxytocin, and you don't have to be a nursing baby to accomplish that. Nursing, like intercourse, has been used to induce labor since prehistory. In some ancient cultures, midwives placed a breast-feeding baby at the nipple of a woman struggling to give birth to hurry things along. Since then, suction cups, breast massage, and even rubbing the nipples with cotton soaked in paraffin have all been used. In 1973, Israeli doctors experimented with a regular breast pump, operated by the test women themselves, and found that it induced labor 69 percent of the time.

There's a direct neural connection between the nipples and the oxytocinergic neurons in the brain. When Barry Komisaruk, of Rutgers University, and Stuart Brody imaged the brains of women stimulating themselves in various ways, they found that stimulating the nipples activated the brain in much the same way as stimulating the cervix. They speculated that, as it does in rodents, this kind of stimulation releases oxytocin from the PVN.

This breast-oxytocin link would explain why men are the only male mammals fascinated by breasts, and women are the only female mammals whose breasts remain enlarged, even when they're not nursing. Humans are the only animal for whom breasts have become a secondary female sex characteristic.

Further adding to the evidence that human mating exploits the mother-infant bonding circuit to create love, neuroscientists at University College London asked women to climb into fMRI machines and, as in other studies we've discussed, look at pictures of their babies. The resulting images looked like other such tests. The same areas lit up, reconfirming the existence of a mother-infant bonding circuit. Then the scientists asked the women to look at more pictures, including likenesses of George Clooney, anonymous strangers, relatives, and their lovers. Looking at Clooney wasn't like looking at their babies—nor was looking at strangers, or even at relatives. But when the women looked at pictures of their lovers, the activation patterns in their brains overlapped uncannily with the patterns formed when they looked at their babies.

Nobody can attempt the kinds of experiments on humans that Larry and his colleagues have run on prairie voles. So the evidence we've presented here isn't definitive, scientific proof: it forms a hypothesis. But when a prairie vole mates, a set of neurons that travel from the hypothalamus to the pituitary fires. The fibers that have drifted from the central bundle leading to the pituitary, and found their way into the nucleus accumbens, fire, too. The fMRI images show that women looking at their babies and their lovers use the same brain structures; these are the same areas, and the same neurons, that control maternal behavior in both voles and people.

If Larry's theory is correct, one might ask why the actors who work for Joy King have trouble looking into each other's eyes. Why don't they fall in love every time they perform? Why did William Butler Yeats write one of his most famous poems, "No Second Troy" (which begins "Why should I blame her that she filled my days/With misery . . . ?") after Gonne resumed her former distance from him? Why doesn't vacation sex always lead to loving monogamy?

As we've stated, context is important: a porn-film set isn't exactly romantic, or even very sexy. More important, as Pfaus's rats show, developing a fetish requires some practice. In its doorman role, oxytocin simply opens the portal to love. One has to walk through it a few times, and under the right conditions. Even then, it has to be constantly renewed. A one-night stand might or might not make you think you're in love, but any effects it's had will wear off.

Human sex isn't only about reproduction and passing on our genes. Most mammals become sexually receptive only when they're fertile. But humans have sex even when there's no chance of fertilizing an egg. Sex, then, must serve some other purpose aside from making more people. We believe that during human evolution the mother-infant bonding mechanism, shared across all mammalian species, has been tweaked so that women can use sex to establish a bond or to maintain one.

Put another way, men are using their penises and their partners' breasts to entice women to babysit them. (A woman may wish to keep this in mind the next time a man uses his mouth to make motorboat noises in her cleavage.) Women want something, too, of course, and, as you're about to learn, they also exploit an ancient male circuit to get it.

6

BE MY TERRITORY

Just as Markus Heinrichs did before he conducted his own experiments, you, too, may doubt that one molecule, or even a series of molecules, could really be at the root of something as complex as human love. The idea would seem to diminish the role of free will in perhaps the most important decision of our lives. We choose when to have sex. We choose whom we'll have sex with and we choose whom to love. That's what most people believe. But while it's not our intention to imply that free will has *no* part to play, we're hoping to show that humans are very strongly influenced by neurochemicals, that human love really is the result of these chemicals acting on defined circuits in our heads, and that the "how" of love—even our ability to truly commit to monogamous love—varies, by individual, according to genetic and environmental events over which we have little control.

To see just how powerful a role neurochemicals have, and for how long they've had it, we've come to Kathy French's lab at the University of California at San Diego, high on a mesa overlooking the Pacific Ocean. French studies leeches. She's spent years studying them, which makes her sound a little eccentric. But she's not at all the way one might imagine a bloodsucking leech aficionado: dark, gothic, misanthropic. She's a perky, petite, energetic blonde who could have been a high school cheerleader.

When French and postdoctoral fellow Krista Todd take us into a tiny, steamy room—a big closet, really—where the leeches are kept, there's not much to see except small glass aquariums sitting on a series of shelves. Inside each tank, along with a little water and some moss, a few leeches are doing, well, nothing. If these leeches could be said to have an attitude, you'd have to say theirs is spoiled arrogance. It would be easy to mistake them for pampered, shelled escargots. They're about three inches long when scrunched up (which they are, most of the time, though they can stretch to eight inches or so). Unless you poke them, all they do—between slurping the occasional ration of blood—is lounge, motionless, by the edge of the water like Jabba the Hutt on vacation in Aruba.

Yet French can recite facts about them with the excitement of a thirteen-year-old girl talking about Justin Bieber's hair. *Hirudo medicinalis*, the medicinal leech, is part of a family of animals called annelids. Earthworms are annelids, and, in fact, leeches are themselves worms. Nobody knows just how evolutionarily ancient leeches are, but they're very, very old. Fossil annelids have been found in rocks that are roughly 500 million years old, from the Cambrian period. *Hirudo* probably aren't quite that old, but the lounging leeches in French's tanks are living relics whose predecessors were lying around in much the same manner, near some swampy pond, long before voles or sheep or people—or any mammal—walked the earth.

All *Hirudo* are hermaphrodites; each one has both male and female gonads. That might seem pretty convenient, but despite having the equipment of both sexes, they can't self-fertilize. In fact, leech sex is downright awkward.

First, a leech has to find a partner. Because they don't have eyes, French believes they "taste" the environment for nearby candidates, probably using chemosensors on their lips. What they're tasting for, she thinks, is urine, released through pores on the underside of leech bodies. When a leech is in the mood, it raises its head like a cobra and smacks its lips open and closed like an old, toothless woman gumming the air, hoping to find some tasty urine and signs in that urine that somebody's ready to mate. They're seeking social information about receptive partners. French can't yet prove this, but, she says, "they must have some sense of it, or else they'd wind up raping each other all the time."

Once a leech finds a mate, it has to engage in some elaborate choreography. More pores, called gonopores, also on the underside of the leech's body, lead to sex organs. The penis is located in a gonopore on the fifth body segment. A female gonopore is located just one segment down, on the sixth. So, if you're a leech, you have to make sure your fifth body segment lines up with your mate's sixth body segment, and vice versa—which isn't easy. To manage it, the leeches have to reveal the undersides of their bodies, and then screw. We mean this literally, as in twisting around each other in what French and Todd call "telephone cording" for the obvious reason that, when leeches do it, they look like squirming telephone cords.

It's hard to imagine lazy leeches that normally lounge poolside doing anything like this. French believes they only reveal the undersides of their bodies, only perform the cobra moves, only do the telephone cording when they're trying to mate, and at no other time. There's a good reason why they should want to shy away from such extravagant behavior: It's very dangerous. When the leeches are screwing, they're vulnerable to predators.

Whatever triggers this has to be awfully powerful to make leeches disregard their own safety. But, though *Hirudo* have been used in medicine for centuries and have gained an even larger role in modern medicine, nobody knew what made them start screwing until French heard a University of Utah scientist speak at a meeting. The scientist, Baldomero Olivera, talked about the venom cone snails use to hunt prey, usually some kind of sea worm—another annelid. The cone snail's venom contains neurotoxins. French, seeing an opportunity, said, "*Hmmm.* We've got annelid worms, and you have these neuromuscular agents. We'd love to check them out on our animals."

The two labs began collaborating. The Utah people distilled the cone snails' venom into its constituent chemicals and sent some of those ingredients to the San Diego people who injected each one into leeches, waited to see what happened, and then reported back to Utah. Both labs figured the testing would go on for many months, if not years. Almost right away, though, French had a shock.

"It took only three iterations before we had discovered that one subfraction of the venom produced all the reproductive behaviors" of the leech—the cobra gestures, the telephone cording—even when there wasn't a partner in the area, she recalls. That subfraction turned out to be commonly available from chemical supply companies. So French bought some. When she shot it into leeches, they performed what she now calls "fictive flirting." They started trying to mate. (Todd had suggested "fictive fucking," which is actually closer to what's happening, but French decided that term might not look good on posters displayed at science meetings.)

To show us how powerful and immediate the effect of the chemical is, Todd takes a thin syringe and holds the needle above a leech sitting alone in a plastic container with just a little water. Stretched out, this leech would be about five inches long, but, as usual, it's not stretched out: it's doing its leech thing and just lying there. Todd injects the leech, and within about two minutes it anchors its tail to the bottom of the container and begins stretching and twisting the rest of its body to expose its underside, opening its mouth and acting very much like a living telephone cord. Eventually, it begins to release the tip of its penis from the male gonopore.

Todd pokes the leech with the syringe needle. Normally, a leech would scurry away. But this one's oblivious to pain; it just keeps screwing. French's lab has even jolted telephone-cording leeches with static electricity, and they won't budge. They are completely taken over by this mating response. They won't do anything but try to have sex. It's possible that the cone snail's venom has evolved to contain a chemical that will hijack the sea worms' mating response, allowing the cone snail to transform the worms into helpless snacks.

That chemical, the same one that turned French's leeches on, is called conopressin. The leech has an analogue of it called annetocin. Humans also have an analogue—two, actually: oxytocin and vasopressin.

Human oxytocin and vasopressin trace their lineage back 700 million years or so. At some point during evolution, the gene for the chemical parent of conopressin and annetocin split into two very closely related genes, eventually becoming the ones for vasopressin and oxytocin. All these chemicals consist of a chain of nine amino acids. Here, for example, is the annetocin chain:

Cys•Phe•Val•Arg•**Asn**•**Cys**•**Pro**•Thr•**Gly.**

Now here's the conopressin chain:

Cys•Phe/Ile•Ile•Arg•**Asn**•**Cys**•**Pro**•Lys/Arg•**Gly.**

This is human oxytocin:

Cys•Tyr•Ile•Gln•**Asn**•**Cys**•**Pro**•Leu•**Gly.**

Finally, here's human vasopressin:

Cys•Tyr•Phe•Gln•**Asn**•**Cys**•**Pro**•Arg•**Gly.**

The letters stand for names of amino acids and their positions in the chains, but don't worry about the names. The letters in boldface type are the ones animals that evolved hundreds of millions of years ago have in common with human beings. In other words, these chemicals have been extremely highly conserved, from ancient leeches all the way to us. In fact, if you were to take the gene encoding the blowfish oxytocin analogue, isotocin, and put it into a rat's genome, the rat hypothalamus would make isotocin within its oxytocin-making neurons. Which means that, despite some differences, the DNA sequences in the fish and mammal genes that provide instructions for where those genes are to be used—the "promoter" sequences—are so conserved that they work the same way in very different species. Also note that human oxytocin and human vasopressin differ only at the third and eighth positions. The two hormones have so much in common, they can bind to, and activate, each other's receptors.

When French's lab wanted to test how an antagonist, a receptor blocker, would compromise the effect of conopressin in the leeches, someone suggested they simply purchase a vasopressin antagonist. "I said, 'Look, these proteins are really complicated and specific. Well, sure, go ahead and buy 'em, but it's gonna fail, so don't be depressed,'" she recalls with a laugh. The antagonist worked perfectly.

Something so conserved must play a very basic biological role, or it would have been tossed onto evolution's roadside long ago. We've already discussed some of oxytocin's roles. It was first discovered in 1906 by Sir Henry Dale, who went on to win a Nobel Prize in 1936 for his codiscovery that nerves talk to each other with chemicals and not with electricity, as most people thought. Oxytocin was re-created in a lab by an American scientist, Vincent du Vigneaud, in 1953. His lab was the first to isolate vasopressin and synthesize it. He also won a Nobel Prize.

By the time du Vigneaud isolated vasopressin, science was already referring to it as the antidiuretic hormone because it's responsible for maintaining the correct water balance in our bodies. Children who chronically wet their bed are often prescribed a vasopressin drug. Back in the 1950s, no one suspected it was also working in our brains to affect behavior. But it may well be crucial to human love, especially from the male point of view, just as it is to leech mating.

Leeches and fluid balance would seem to be a world away from monogamous human love. But let's say you're in the basement, trying to nurse your furnace back to health, a chore made a little more tolerable because you've brought a bottle of beer down there with you. Then, after you realize you have no idea how to fix a furnace, you feel a thirst coming on and reach for the beer. Now you realize the opener is way upstairs, in the kitchen drawer. Instead of climbing up there to retrieve it—facing uncomfortable questions about the furnace in the process—and instead of creating a bottle opener on the spot, you reach for a handy screwdriver and find that it works pretty well. Evolution also adapts existing tools to new uses. Vasopressin and oxytocin are examples of what evolutionary biologists call exaptation: using a preexisting molecule or circuit for a new purpose.

Leeches pee for the same reason people do: to get rid of extra water and the waste it contains. But they can also get a lot of information from the urine of other leeches, like if one nearby is ready to screw. Mammals use pee all the time for communicating—just ask your dog. When you take Sparky for a walk, he's likely to spend an inordinate amount of time spritzing a little pee on his favorite landmarks in your (or, rather, his) neighborhood. He's sending messages. Mammals can tell a lot about others through their urine, including, for some species, who might be willing to have sex. Many species mark their territory with urine squirts, telling the world that the space they've marked out is theirs.

Some don't use urine to mark territory; they use scent glands. In 1978, and again in 1984 (it more or less had to be rediscovered), scientists found that when they gave male hamsters injections of vasopressin into the medial preoptic area (MPOA) and the anterior hypothalamus, the hamsters went into a frenzy of turf grabbing, using the scent glands on their hindquarters to mark out territory like homesteading Okies during a land rush.

Frogs don't use urine or scent glands; they talk to each other. Male frogs advertise their availability to females and, in some species, establish territory, by croaking. Croaking is set off by the action of vasotocin—their analogue of vasopressin—in frogs' brains. There's evidence that croaking is modulated by water retention, which increases pressure inside the frog and makes for a decent bellow. So, for them, vasotocin is vital to mating-related communication and marking territory.

In order to claim territory successfully and effectively, of course, an animal has to remember where its territory is. This requires spatial memory—akin to the social memory we explored in the previous chapter—so it can distinguish its territory from that of all others.

Also, if you're going to go around claiming territory as your own, there's a good chance you're going to have to defend it. So you have to be willing to fight, or what's the point of claiming it in the first place? And, if you're willing to defend it, risking pain, injury, maybe even death, then you must be pretty attached to it. You could even be said to be bonded to your territory. You have a home. No other place is your home, just this place. You've become, so to speak, territorially monogamous. When you mate, chances are you're going to either invite your new partner into your territory or establish some new territory as your own—a little piece of ground where you can raise the kids—and you're going to zealously guard your mate the way you'd zealously guard your territory.

What started as a way to control water balance has now created an affinity in human males for bonding with a mate. Just as love has its evolutionary and neuronal roots in the mother-infant bonding circuit for female humans, for male humans, Larry believes, females are territory.

As we've said before, this is a hypothesis, not a proven scientific fact. So let's stipulate right now that both oxytocin and vasopressin are present and serve physiological and neurosignaling functions in both men and women. Science hasn't completely figured out how they may interact to influence mating behaviors in people. But we do know quite a bit from animal experiments.

It's now established that, while oxytocin sensitivity is more dependent upon estrogen, vasopressin production in neurons projecting into behavior-controlling regions of the brain depends more on testosterone. Males have higher levels of vasopressin-emitting neurons in the amygdala, and vasopressin does seem to regulate macho social behaviors, like mate guarding, in a variety of species.

True, if you shot vasopressin into a man, he wouldn't immediately drop the pizza and start telephone cording. So we ask Kathy French if she thinks it's a stretch to compare leech mating to human mating.

"It's remarkably parallel to humans," French replies. "I think, right down to the molecule."

Todd nods with certainty. "Right down to the molecule," she adds.

THE GIRLFRIEND CHEMICAL

For those species in which it exists, monogamy is important—even for, and sometimes especially for, males. The male angler fish, a deep-sea denizen, takes monogamy very seriously. Anglers live so far under the surface that light is virtually shut out of their world. So these fish have evolved a lantern, a little bioluminescent bulb-on-a-pole they use to attract prey, and, perhaps, to find each other. Even with the aid of the lantern, though, it can be tough to find a mate that far down. So when they stumble upon a female, male anglers form an actual, not just metaphorical, bond. They bite into her, fuse blood vessels, and then dissolve away until the males exist mainly as a hypothalamus and sack of testicles attached to a female's body.

Female anglers aren't as loyal. They can be found wearing several sets of testicles, like so many trophies. Don't feel sorry for the male, though. He's gotten what he wanted most, evolutionarily speaking. Every time the female spawns, he'll produce some offspring, dropping into the angler population whatever genes made him select this lifestyle. The primordial angler males whose brains weren't organized to pursue this extreme monogamy were unlikely to produce many offspring at all, given the environment. As a result, those bachelor genes have largely faded away from the gene pool. That's how natural selection works. Behaviors adapt to environments to produce the most surviving offspring.

In 1993, James Winslow, Sue Carter, Thomas Insel, and others announced they'd discovered that vasopressin activity in the brain played a role in mammalian monogamy. They had conducted a series of experiments with prairie voles. Before mating with a female, the males happily associated with other voles of both genders. If a strange virgin male arrived in another virgin male's cage, the two would sniff and investigate each other, and that would be the end of it. All that changed once the males mated, however. They developed a partner preference for the female they'd mated with, and then started attacking any strange vole who came into the cage.

Vasopressin wasn't as obvious a candidate for a role in bonding as oxytocin: science already knew that oxytocin was involved in affiliative behavior. By the time the Winslow experiments were over, though, the scientists had proved that brain vasopressin, which is released in males' brains during mating, not only was involved in their postmating behaviors, but the male prairie voles wouldn't do the behaviors without it. If vasopressin was blocked, they didn't form a partner preference even if they mated. Without vasopressin, males have very poor social memory. And while they would mate if vasopressin was blocked, they wouldn't act aggressively toward other males afterward.

They also found that if they infused vasopressin into a male's brain while he was with a female for just a few hours—*without* mating—he would prefer to be with that female over others, even if she ignored him. Subsequent research showed that mating, and then living with a female, physically changed the male brain, increasing the density of vasopressin-emitting nerves and reorganizing the nucleus accumbens. Both changes strengthened the bond and stimulated paternal care for young.

Male meadow and montane voles mate just like prairie voles do, and they, too, get a shot of vasopressin. The relevant neurons, and their projections, look the same. But they don't bond with their female partner, and they don't show any increase in aggression toward intruding males. One might assume that the meadow and montane voles must be getting too small a dose of vasopressin to bond. But Insel found it's not a matter of dosage.

Rather, like the oxytocin system in females, it isn't the *amount* of the chemical but the specific brain regions that are *sensitive* to it that counts. This differs among the species. One region, the ventral pallidum (the major output of the nucleus accumbens—Figure 2, page 76), is packed with vasopressin receptors in male prairie voles—but not in meadow voles. The other region is the lateral septum, where vasopressin receptors are necessary for a male to remember a particular female. When Larry and his colleague Zuoxin Wang blocked vasopressin receptors in either of these places, a night with a sexy female was no longer enough to create a bond. So Larry believes it's this distribution of receptors that drives male prairie voles to bond with females.

Six years after Winslow's groundbreaking experiments, Larry ran a very simple, but revealing, test. He injected either vasopressin or a placebo into the brains of male prairie and montane voles, then placed them in one side of a two-sided arena with an anesthetized female. All the montane voles approached and sniffed the sleeping female, figuratively shrugged their shoulders, and then wandered around, exploring the cage. It didn't matter if they'd received the vasopressin or not. But prairie voles injected with vasopressin sniffed the female, nuzzled her, and groomed her much more than the prairie voles given the placebo did. It's a bit like French's fictive flirting, except that the voles had a real female—albeit an unconscious one—to flirt with. This may sound simple, but the implications are tremendous. It proves that the same vasopressin in the brains of these two very similar species have very different effects on the social behavior of each.

Receptors are proteins, and proteins are encoded by genes. Genes have different parts: the coding sequence (or recipe) and the promoter. The promoter determines which cells will make the protein—in this case, the receptor for vasopressin. Larry wondered if differences in the promoter of the vasopressin-receptor gene, called *avpr1a*, might account for the differences in the pattern of expression in the brains of prairie, meadow, and montane voles. So he isolated the *avpr1a* gene from prairie and meadow voles and sequenced the DNA.

The genes were 99 percent identical in both species, suggesting that the structure of the receptor protein is identical. But there was a difference in part of the promoter, in a stretch that geneticists often refer to as "junk DNA" because it was once thought to serve no role but to hang there, like junk. Because this junk is made up of repeated sections, it can sound like a broken record to a cell's DNA-replication machinery, causing the machinery to stutter and leading it to spit out more, or fewer, repeats than exist in the sequence it's trying to copy. In other words, junk DNA, more properly called microsatellites, is a hot spot for genetic evolution. (In fact, microsatellites can be used as a genetic fingerprint, à la TV's *CSI*, to identify a suspect.) This led Larry to believe that microsatellites are responsible for diversity in receptor distribution in the brain and, therefore, for diversity in behavior.

But does the receptor distribution in the brain really determine behavior? To figure that out, Larry and his team placed the *avpr1a* gene, with its junk DNA, promoter, and coding sequence, into mouse embryos to create transgenic mice with brain vasopressin-receptor patterns that looked like a prairie vole's. When those male mice—which are firmly polygamous—grew up and were injected with vasopressin, they acted a lot more like prairie vole males when they encountered female mice than their normal brothers did. They did much more sniffing and grooming of the females.

Nothing else in those mice was different. Oxytocin-receptor expression was the same for both the transgenic and normal mice. They walked around and explored in the same ways, and, when Larry gave the mice something else to smell—cotton balls doused with a lemon scent or with the scent of female mice who'd had their ovaries removed—there still wasn't any difference. The only behavioral change between male mice with a prairie vole vasopressin-receptor gene and male mice without one involved how the males related to an actual female. The transgenic males were more prodigious flirters. This experiment was one of the first to prove that mutations in a regulatory-promoter region can profoundly affect behavior. The implication, of course, is that behavior, something we often regard as fixed, at least when we're talking about species-typical behavior, can actually be subject to highly changeable bits of molecular string.

Because the mice didn't become monogamous like prairie voles, a natural question was whether *avpr1a* from a prairie vole placed in a more closely related promiscuous vole would do the trick. So Larry's lab conducted a similar series of experiments in male meadow voles.

They focused on the dopamine-reward circuit—specifically, on the ventral pallidum, the region where Insel found differences in receptor density among the species. (Monogamous field mice and marmosets also have loads more receptors in their ventral pallidums compared to closely related nonmonogamous mice and monkeys.) Larry harnessed the power viruses have to infect a cell and inject their own viral genes into it. He deleted a virus's genes and replaced them with the prairie vole *avpr1a* gene. Then Miranda Lim, his student, injected that virus into the ventral pallidum of meadow voles. The virus did its job, and the cells in the meadow voles' pallidum began making vasopressin receptors at about the same rate as male prairie voles normally do.

The males were then put in a cage with a receptive female for twenty-four hours. After that, the voles were given the partner-preference test. As expected, the couples had all-night sex, just as normal meadow voles do. Later, when given the choice between the female they'd spent time living with or a different one, control meadow vole males that had been injected with a placebo virus didn't show any preference. They investigated and huddled with the stranger as often, and for as much time, as with the girlfriend—like meadow voles are supposed to do. The meadow voles with the prairie vole *avpr1a* gene, however, spent much more time huddling with the girlfriend, as a prairie vole would.

Larry's team had switched on monogamy. Or, if you prefer, they'd created a pair-bond in a species that normally doesn't pair-bond. They'd changed an innate behavior with one genetic alteration, and it wasn't even an entirely new gene: it was merely a different version of a gene the meadow voles already possessed. Not only did that slight variation make a big difference between one vole and another vole, it also created a different social system: monogamy versus polygamy. That's a very thin curtain dividing profoundly different ways of life.

The explanation for why the experiment worked is much like the explanation for female bonding in the last chapter (and, in fact, oxytocin and vasopressin nerve fibers can be found in parallel in those parts of the brain where they exist). Motivated by androgens acting on their heterosexually organized male brains, and by appetitive reward piqued by the smell of an estrous female who hopped and darted, the males mated and got the brain reward. During sex, they took in the females' aromas and any other social information they could glean. Sex also produced a surge of vasopressin from fibers originating in the amygdala. As the nucleus accumbens and ventral pallidum were being bathed in dopamine, the social signature of the partner was converging on the lateral septum and ventral pallidum. The convergence of dopamine, along with these social cues, resulted in strong neural connections that linked the cues with the reward.

Now the meadow voles, like their male prairie vole cousins, could associate an individual female with the reward. Having a girlfriend suddenly seemed a like wonderful idea.

Scientists have since learned that the curtain can be torn, that the junk DNA responsible for this difference is changeable, even within a species. In fact, when Larry was studying the repeat stretches of microsatellites, he noticed that not all prairie voles had the same sequence. There was quite a range in the length of repeated sections among individuals. There was also lots of individual variation in social behavior among male prairie voles. In nature, about 60 percent of males will settle down with a female partner. The rest play the field.

Elizabeth Hammock, a former student of Larry's, who is now at Vanderbilt University, wanted to know whether the variations in the length of the *avpr1a* microsatellite might be related to variations in vasopressin-receptor patterns, and in social behavior, among prairie voles, just as such variations exist between prairie voles and meadow voles. She screened all the voles in the colony, found males and females with long and short versions of the microsatellite, and then played matchmaker by breeding long females with long males and short females with short males to create two groups of prairie vole babies: those with long microsatellites and those with short ones. Then she fostered half the resulting pups to mothers who had the opposite version, so any differences in rearing would be minimized and any future behavior would be more likely to result from the gene variant—not from Mom's licking and grooming.

The males who'd been bred for the long version of the *avpr1a* microsatellite had a higher density of vasopressin receptors in several brain regions, including the olfactory bulb and the lateral septum (good for social memory), and were accomplished lickers and groomers, showing lots of parental attention to their babies. New fathers with the short version of *avpr1a* performed the licking and grooming behaviors significantly less—up to 20 percent less—almost matching the degree of difference Frances Champagne sees between her high- and low-LG rat mothers. There was no gene-length-version difference among females, showing that the response to the vasopressin-receptor gene variant is male-specific.

The different versions of the gene led to different paternal styles. But what about mating and bonding? Hammock stuffed soiled bedding from the cage of a similarly aged female into the males' cages. The males with the long version of *avpr1a* were quicker to investigate the aroma, did so more often, and hung around the smelly bedding longer than the males with the short version did. These differences were seen only in response to the female aroma, and not when Hammock tried other scents, like banana. Most important, when the males were paired up with a sexually receptive female for an overnight visit, the long-version males, as a group, spent at least twice as much time with their new girlfriends during a later partner-preference test. The short-version males showed little or no preference. They were crummy bonders.

These experiments showed that variations in the length of so-called junk DNA can have very profound effects on behavior—at least in the well-controlled confines of a shoe-box-sized vole apartment. It's true that other studies conducted in wild, or wildlike, environments have produced conflicting results. One such experiment, led by Alex Ophir, of Oklahoma State University, found it wasn't necessarily the microsatellites that determined how faithful a vole would be. Rather, the pattern of the receptor distribution in its brain predicted how many babies a male would have, and with how many different females. Prairie vole males with few vasopressin receptors in the cingulate cortex, which processes spatial memory, seemed to "forget" their territory and became "wanderers," mating with multiple females and having more pups.

Still, there's no doubt that vasopressin-gene expression has profound effects on male vole mating behavior, as well as parenting, and that gene variation affects those behaviors.

To place such findings in a larger context, Larry searched through gene databases to see if other mammals possess similar diversity. When he looked at the *avpr1a* of Clint, the first chimpanzee to have his genome sequenced, Larry found that a huge chunk of DNA in the microsatellite, RS3 (repetitive sequence 3), which, in humans, is variable in length, was completely missing. Interestingly, chimps are infamous for violence, infanticide, and the sexual harassment of females.

Puzzled, he and a student, Zoe Donaldson, looked at the region in eight more chimps and found that about half had an *avpr1a* RS3 microsatellite that looked very similar to humans. The other half were like Clint; they didn't have it at all.

William "Bill" Hopkins, a psychobiologist at the Yerkes National Primate Research Center, and one of Larry's colleagues, has discovered that chimp personality traits, like dominance and conscientiousness, are linked to RS3 variation. Male chimps that possess two copies—one from Mother and one from Father—of a short version of RS3 display significantly more dominance behavior and less agreeability. They also have less gray matter in the anterior cingulate cortex, the same area implicated in vole territory forgetfulness. The anterior cingulate cortex is interconnected with the prefrontal cortex (PFC), which we encountered before when we discussed how reward hushes critical thinking.

Bonobos are so closely related to chimps that, until 1929, biologists thought they were the same species. Yet, they're a much less violent group and, while hardly monogamous, do form social bonds, often using sex as the glue. The bonobo *avpr1a* microsatellite region studied by Hammock is almost identical to the same human *AVPR1A* region. (Human gene names are capitalized.)

Nobody can yet say for sure if this genetic similarity really is linked to the bonobo social system. But there is tantalizing evidence that this very RS3 region does influence vasopressin-receptor expression in the human brain, and predicts behavior, just as one might expect from Larry's vole studies.

MATE GUARDING, OR, WHY HE WANTS TO KICK YOUR ASS

Prairie vole males will defend their nests and aggressively guard their mates against any intruders, especially other males. When an intruder does arrive, brain vasopressin in prairie vole males shoots up by nearly 300 percent. When hamsters smell an intruder, they start vigorously scent marking, just as they would if a scientist shot vasopressin into their brains.

Mammals of all kinds, including people, are given to such mating-related displays of dominance and guarding around other males, including people.

Sean Mulcahy starts his story the way lots of guys who've gotten into trouble start a story: "I met this girl."

He's thirty-one at the time we talk to him, a good-looking guy—and big—with a brown mustache and goatee and longish sideburns. He runs a muffler shop. He's the kind of guy who likes to party, likes fast motorcycles, and doesn't shy away from physically asserting himself. Raised in an Irish Catholic working-class family outside of Chicago, he and his brother spent a good part of their youth doing what a lot of Irish Catholic brothers do: fighting with each other. Still, Mulcahy considered his brother his best friend.

In 2001, Mulcahy was arrested for damaging property. He was leaving a bar with his then wife when he mouthed off to another man. "The guy was driving a convertible, and I said something extremely stupid; I'm the one who pissed him off," Mulcahy admits. "I said something like, 'Guys do not drive convertibles, girls do, and the guys who do squat to pee.' I was twenty-one; you know: a guy being an asshole. I didn't think he was gonna bust my window out!"

Sean Mulcahy was, in his way, communicating socially. He was mate guarding and trying to establish his dominance over the convertible-driving dude. One could also say he was being a jerk; he's not shy about saying so, nor is he shy about displaying his regret over the incident. But Mulcahy is not, by any means, a full-time jerk. He's a hardworking man who originally trained as a millwright. He weeps freely. He's close to his mother, like the good Irish Catholic son he is. And when he talks about love, he seems possessed by it.

Mulcahy and his first wife had been divorced for years when he met his fiancée. He'd promised himself never to marry, or to trust, another woman, and he'd kept that promise. But, then, "it was so weird," he says. "Within a week of meeting her, she was everything to me. Everything I'd been on guard about just went out the window. I felt so strongly for her. We had a real bond, a serious connection."

At the time, Mulcahy and his brother shared a house—their parents' former home. The house had a pool. Mulcahy's fiancée would come over, and so would one of her girlfriends, and it's fair to say that alcohol was sometimes consumed. One day in August 2010, when Mulcahy's girlfriend was in the pool with her friend, Mulcahy's brother jumped in.

This is where stories begin to diverge, and we're not going to try to sift through the accusations, counteraccusations, and defenses here. Suffice it to say that Sean Mulcahy came to believe that his brother had sexually taken advantage of his fiancée and that he felt outrageously aggrieved when he learned of it. "This was a superintense betrayal," he says.

Three days later, after tormented sleepless nights and desperate calls to his mother, Mulcahy confronted his brother, who was staying with a friend. "This was my woman, my fiancée, and especially . . ." He pauses for a moment, not completing the thought. "So I just wanted to punch him in the face as hard as I could." But, using his prefrontal cortex, he reasoned that one punch wouldn't be the end of it. "I'm about two twenty-five, he's about two thirty-five, and we're both about six feet four. If it got physical, it wouldn't stop. That scared me." Somebody could wind up dead, he figured. As strong as the urge to fight his brother was, the possible consequences were so enormous that Mulcahy withdrew. He left, returned home, and began drinking the unlikely combination of Hi-C and vodka. He sat on the front steps and "bawled."

Mulcahy usually carried a work-related utility knife on his belt. When his brother's cat approached him, and Mulcahy shoved it away, the cat retaliated by scratching him. In an instant, Mulcahy seized his knife and slit the cat's throat. He immediately regretted this. He took a picture of the dead, bloody animal with his cell phone and e-mailed it to his brother with a message: "This is what you did to me." It was about 4:30 a.m.

He'd backed away from a fight earlier. But, this time, he didn't engage his rational brain at all, which was, in any case, crippled by Hi-C and vodka. "I was definitely out of control," a contrite Mulcahy says. "It wasn't premeditated. I couldn't believe I'd even done it. I called my mom right away, and said, 'I think I need to admit myself.' It didn't even seem real."

In court, he pleaded guilty and apologized profusely. In March 2011, the judge sentenced him to a long probation with community service. He lost tens of thousands of dollars in the fallout and received death threats from animal-rights agitators.

The justice system, and society at large, punishes this sort of behavior, of course, and rightly so. Mulcahy himself acknowledges he got what he deserved from the judge.

About three thousand years ago, we had a different attitude toward retaliating against male intruders. "The unconquerable Odysseus looked down on them with a scowl," Homer wrote in the denouement of the returning hero's story. " 'You curs!' he cried. 'You never thought to see me back from Troy. So you ate me out of house and home; you raped my maids; you wooed my wife on the sly though I was alive—with no more fear of the gods in heaven than of the human vengeance that might come. I tell you, one and all, your doom is sealed.' "

And then, in an orgy of slaughter, Odysseus killed a lot of people.

We may have become more "civilized" since then, but you can't civilize our biology out of us. Accounts of bar fights, stabbings, and shootings over women pepper the criminal calendars of every court. Like that of the ancients, great modern literature has turned on this theme. When Tom Buchanan confronts Jay Gatsby about the mysterious Gatsby's intentions toward Buchanan's wife, Daisy, he asks, "What kind of a row are you trying to cause in my house anyhow?" And soon, out of revenge, and to mate guard Daisy, Tom falsely implicates Gatsby in a death, a lie that ultimately leads to Gatsby's murder.

My house. *My* wife. *My* girlfriend. *My* home. Like Buchanan, men often use words like "home" and "wife" interchangeably. When nations have seen fit, they've even tried to tap into this ancient urge with propaganda posters showing a caricature of an enemy reaching not so often for land but for a woman, who represents all wives and girlfriends. Enlist to protect her!

We may couch this urge in patriotic fervor or heroic poetry, but we're still doing what the lowly prairie vole would do. Mulcahy wouldn't have killed the cat if his brother had been linked to some other woman; he did it because his brother was linked to *his* woman. A guy doesn't get into a bar fight over a woman he's not dating or in love with; he gets into a fight because a man is flirting with *his* woman. The Greeks had no compunction about raping and killing the women of the conquered, but Odysseus was outraged to think men were wooing his own wife. Mating makes men behave differently toward those who intrude on their female.

In Sean Mulcahy's own words, he had a strong "bond" with his fiancée. (No, we didn't suggest that term to him—he came up with it on his own.) Driven by hormones and appetitive reward, sex released vasopressin (and oxytocin) and led to consummatory reward and the powerful salience of his girlfriend. Or, as most people would say, he fell in love. We cannot say this with scientific certainty, of course, because we can't do to Mulcahy what neuroscientists can do to rodents. Neither can we say with certainty what molecular events occurred in Mulcahy's brain to bring on his act of "catricide," no "vasopressin made him do it!" But just as the story of oxytocin and female love is being pieced together for women, there's emerging evidence in humans that voles and even leeches really are good analogies for how men mate and form love bonds.

Vasopressin biases the brain toward sex. Kathy French sees it clearly in her leeches. She once tried to blind herself to whether or not leeches received an injection of conopressin so she could objectively score the leeches' behavior, then see if there was any difference between the conopressin and a placebo. She couldn't blind herself, though, because the conopressin-induced fictive flirting was so obvious. In Todd's demonstration for us, the leech was alone, so it stood there twisting all by itself: fictive masturbation. But when French was trying to compare the vasopressin with the placebo, she was testing behavior in the presence of other leeches.

"One leech came out, and it would chase every other leech around the dish," French recalls. "I thought of the icky guy in the bar hitting on every female. It was just like that! The others would go, '*Eww!*' and move away." The injected leech would taste the air for signs of others, trying to collect social information. But even if he didn't find any good candidates, the vasopressin had made him so single-minded about mating, he kept approaching others anyway. "This guy was, taste, taste, taste, and, every time, the others went, '*Eww, eww, eww,*'" French says with a laugh.

So we know that, in leeches, at least, either annetocin or conopressin creates a single-minded focus on sex. In fact, the nerve that controls the leech penis is studded with annetocin receptors. Men may not become quite so single-minded, but vasopressin is involved in human erection and ejaculation, and, under its influence, our brains are more receptive to sex cues.

Adam Guastella showed this in an experiment with words. Guastella gave either vasopressin or a placebo to men. He then showed them a series of words, in random order. The men who received the vasopressin spray recognized sex-related words more quickly than non-sex-related words. The extra vasopressin increased the valence of the sex words. Spraying the vasopressin up the men's noses created an affinity for sex. And, though it hasn't been tested in humans, Heinrichs believes vasopressin also influences men to judge female faces more attractive than they would without extra vasopressin, in the same way women with more oxytocin regard male faces as more attractive.

Whether or not this affinity is expressed can depend entirely on environment. If you don't show sex words to a man with extra vasopressin sprayed up his nose, he may not think of sex.

For some time, scientists have known that people respond, often unconsciously, to the emotions of others; it's part of our social coping. The reactions aren't only in the brain, though; they can be detected in facial muscles without us having any sense that a muscle has moved at all.

A muscle that scientists often observe via electrical-activity detectors is the corrugator supercilii, a small wedge-shaped one just along the inner eyebrow. Studies have shown that this muscle provides a telltale, if usually unconscious, sign of competitive or aggressive emotions. When Richmond Thompson, a neuroscientist at Bowdoin College, performed one of the first tests of intranasal vasopressin in men, he exposed test subjects to pictures of happy, angry, or neutral faces. A shot of vasopressin up the nose didn't make a difference in reactions to any of the faces, as measured, for example, by heart rate. But Thompson found something strange. Under the sway of vasopressin, the men's corrugator supercilii muscles reacted agonistically to angry faces—which you might expect—but also to neutral ones. Thompson interpreted his results to suggest that vasopressin biased the brain toward seeing neutral expressions as potentially combative.

Thompson, along with colleagues from Harvard Medical School, expanded on this small test. Some men and women were given the intranasal spritz of vasopressin, and some were given a dummy spray of saline. Then he showed each gender a series of angry, happy, and neutral faces. Men looked at male faces; women looked at female ones. The subjects also rated the approachability of the faces they saw, then completed a questionnaire about the levels of anxiety they felt.

Men who received either saline or vasopressin displayed equal corrugator supercilii activity when they looked at both happy and angry faces. In other words, when there wasn't any doubt about what the person in the picture was feeling, both groups of men reacted the same way. Only the vasopressin group produced significantly larger electrical responses to the neutral faces of other men, as if anticipating trouble.

Women had the exact *opposite* reaction. Vasopressin *reduced* the activity of the muscle. In fact, the muscle was significantly inhibited in response to both happy and angry faces.

Vasopressin men rated the happy faces of other men as much less approachable than did men who got the saline spray. Women given vasopressin deemed the neutral faces of other women significantly *more* approachable than women doped with saline did. They also showed a slight uptick in their approachability ratings of happy faces compared with the placebo group. Vasopressin increased anxiety in both genders.

This might seem puzzling. Vasopressin increased anxiety in everyone, but it tended to make women more eager to approach other, strange women. In men, on the other hand, vasopressin not only increased anxiety, it also influenced them to believe that strange men were less approachable. It appeared to bias male brains into anticipating conflict when looking at a neutral male face. It was as if, in the absence of certainty about another man's intentions, vasopressin was telling male brains to assume the worst but telling female brains to reach out to others in times of need.

There are hypotheses about why this might be so. According to one, heightened anxiety in women can prompt what Shelley Taylor, a social neuroscientist at the University of California at Los Angeles, has called "tend and befriend." Women will seek safety with other women, goes the argument, while men respond by preparing for battle or running away: the fight-or-flight response. This could be. It could also be that vasopressin works differently in different parts of the brain. Some research has shown that vasopressin's effects are subtle and varied and that gender-based brain organization and receptor distribution are responsible for these differences.

In 2010, Guastella gave men intranasal vasopressin or a placebo, then showed them happy, angry, and neutral faces. Then he sent them away, asking them to return the following day. When they did, he had them look at a much larger group of photos, with the previous day's images mixed in. He asked them to report whether each photo was "new," "remembered" (meaning it seemed familiar), or "known" (meaning they recalled it from the previous day). The men who took the vasopressin were much less likely to make mistakes by incorrectly identifying new pictures as remembered or known and much more likely to correctly say they knew one of the photos from the previous day. Vasopressin improved their social memory. But when Guastella sifted the data, he found almost all the improvement came from increased recognition of happy or angry faces, not neutral ones. The vasopressin enhanced the men's ability to register clear emotional signals in their memories, just as oxytocin can enhance the salience of social cues.

How might this work? To find out, Caroline Zink, a researcher at the National Institutes of Health, conducted an experiment combining fMRI imaging with a face-matching task. She gave men either intranasal-vasopressin or a placebo spray and then placed them in the machines. Once there, the men were asked to look at a screen and, using left or right buttons, answer which face shown at the bottom of the screen matched a face at the top of the screen. Sometimes the images weren't faces, but shapes such as circles or triangles.

All the men showed an important difference between face and object matching. When they looked at faces, their amygdalas were significantly more active, regardless of whether they took the drug or the phony spray: just as we've seen before, faces evoke amygdala action that generic objects can't.

There was a difference, however, in the signaling between the amygdala and the medial PFC. When we're in social situations that create fear or anger, our amygdala signals the mPFC, which figures out what to make of things, and then sends information back to the amygdala. It's a feedback loop.

In the men who received the placebo, this amygdala-mPFC loop was activated as expected. The amygdala signaled a region in the mPFC, which signaled another, neighboring region in the mPFC, and then that one signaled the amygdala to calm down. The loop is negative: the mPFC tamps down fear and aggression-related activity in the amygdala.

But the communication between the two mPFC regions was broken in the men who received the extra shot of vasopressin. And, because it was broken, there was no message returning to the amygdala to tell it there was nothing to worry about. The amygdalas in the vasopressin men would then be more active for longer in response to the negative faces, making those faces easier to remember.

It's not yet known for sure whether vasopressin works directly in the amygdala or via such feedback circuits. But whatever the details, it seems clear that, in response to negativity, vasopressin encourages wariness and aggression and biases the male brain toward transforming ambiguous social cues into negative ones. The disconnect Zink found in the amygdala-mPFC feedback loop may have the effect of preventing men from thinking too much about taking aggressive action, just as Mulcahy didn't think about killing the cat, and the bias of men given vasopressin toward interpreting neutral male faces as negative may be a hair-trigger survival strategy.

In 2011, Israeli scientists used a test called Reading the Mind in the Eyes. Originally developed by Simon Baron-Cohen, it's become a standard measure of one's ability to empathize with others. During the test, participants are shown a long series of pictures featuring only the eye area of human faces. For each photo, they're asked to select one of four possible emotions, or states of mind, that best matches what they see in the person's eyes. The four choices for each face come from a list of possibilities, some of which can seem a little esoteric. In addition to obvious selections like anger, for example, the menu includes states like "fantasize," "desire," and "despondent." In all, the Israeli test offered ninety-three states of mind and featured both male and female eye regions.

Some men sniffed saline drops, and an equal number took the real vasopressin. Contrary to intranasal oxytocin studies, the men who got the real drug made significantly more errors when they tried to match mental states to the photos. But when the researchers looked at *when* the men made the errors, they found that the entire difference between the placebo and the drug groups was accounted for by the gender of the eyes in the pictures. Men who took vasopressin were lousy at judging the emotions that lay behind the eyes of other men. They did just fine when they looked at women. The placebo men, on the other hand, were *better* at judging men's emotions than women's. When the scientists refined the results even more, they found that it wasn't just the gender in the pictures that mattered but rather gender combined with the particular emotion displayed. The vasopressin effect only applied to images of males displaying negative expressions, like those of upset, accusation, and hostility.

Vasopressin may help you recall negative faces, but if you think you might need to fight somebody, for example, if you're guarding a mate or territory, it's best not to be overly worried about exactly how they're feeling. Too much empathy could get you killed, whether you're a vole, a monkey, or a man.

This leads to the somewhat disturbing conclusion that for men, sex, love, and aggression are inextricably mixed in the brain. It makes sense, in light of Larry's theory that vasopressin's role in territorial and mating behaviors has been adapted by humans so women become an extension of territory in the male brain. If Larry is correct, a man is likely to bond strongly to his mate and to be aggressive in defending her.

Of course, we're not arguing that a woman is literally her man's territory; we're contending that his bond to her engages neural systems that originally evolved for regulating territorial behavior. Neither are we suggesting that this is the *only* component of a man's bond to a woman. But the territorial urge plays an important role.

Aggression is a social act. It informs others that boundaries—personal or physical—should not be crossed. It tells others that "this is mine." But do men like Mulcahy really show this kind of defensive or retaliatory aggression under the influence of vasopressin?

Back in Freiburg, Germany, Bernadette von Dawans, a member of Markus Heinrichs's lab, describes an experiment she just ran using men in an economics game similar to the ones we've already described, but with a twist. "You have to choose to trust or not to trust," she begins. "You have ten Euros, and can either keep all ten, or you can give all ten away to the trustee, where it will be tripled, and so you hope that the trustee, who now has thirty Euros, will return some of that money, like, 'I hope I get eighteen Euros, or twenty Euros'"—a nice bump from one's original ten.

Von Dawans made it attractive to trust by telling subjects that if they didn't want to play the game at all, they could keep the ten Euros and leave, but if they did want to take a risk, they had the chance to triple their money. However, the trustee didn't have to return anything. If you played, he could keep your ten Euros and the "interest" it earned, or he could return, say, a rather insulting five Euros, or the original ten and keep the twenty he earned off of it. "The fair split is to give back half—fifteen Euros—and many do that," she explains. "But there are always some who do not." Thirty Euros is a nice bit of change, so it was tempting for trustees to keep all, or the major part, of their windfall.

Such greediness was exactly the condition von Dawans hoped to achieve so she could then test the twist in the game. If subjects stayed and played the game, they received an allotment of thirty points that could later be translated into cash. They could use these points in only one way: to erase points from trustees. Each point an investor spent deleted three points from the trustee. The investor gained nothing by spending his points this way—the trustee's points did not accrue to him, they just disappeared—so traditional economic theory would suggest that rational investors hoard their points in order to maximize their own profit.

Von Dawans gave her test subjects either a dummy spray, a spray of oxytocin, or a spray of vasopressin, then began the game. Unlike in other experiments, which involve degrees of trust, there wasn't much difference between the oxytocin- and placebo-spray groups under these all-or-nothing rules. Both tended to trust their opposite numbers. So did the vasopressin group. But when trustees returned less than half of the resulting thirty Euros, the men who got the vasopressin were quick to punish the scroogelike trustees. They punished them hard, too—even at the cost of using their own points to delete points from the trustees' accounts.

"This is not the economic way," von Dawans explains. If reason controlled their actions, the investors would pocket all their points, thank their lucky stars for walking away with something, and go buy a beer. But they felt wounded, and perhaps a little diminished, so they lashed out.

Von Dawans puts a positive spin on the matter. Teaching a betrayer a lesson may be harmful to the teacher's self-interest, but it benefits the group. She and Heinrichs both see this aggressive punishing behavior as a metaphor for defending territory. A greedy trustee could be said to "step over the line" of acceptable behavior and injure the group. Punishment establishes boundaries, like a guy in a bar saying, "Don't mess with my woman!"

"Most of us think of testosterone as the hormone that is responsible for male aggressive behavior," Heinrichs argues, "but vasopressin explains this effect much better."

WHY MEN ARE LIKE VOLES

Humans have yet another thing in common with voles: the same kinds of gene variations influence individual behavior. Such variation could explain why some men are the marrying kind, and some men are not, or why some men are better than others at relationships.

People, like voles, carry different versions of the *AVPR1A* gene. The human *AVPR1A* has several microsatellites in the promoter that vary in length. Variation in two of these microsatellite regions, RS1 and RS3, has been linked to variations in brain activity and social behavior. (RS3, recall, is the microsatellite that Bill Hopkins linked to personality traits in chimps.)

Using Larry's work as a prompt, a team from the National Institute of Mental Health recruited hundreds of volunteers to be genetically analyzed and scanned in an fMRI machine while taking the face-matching test that Caroline Zink used to explore the amygdala-mPFC feedback loop, and then to take a personality survey. More than a hundred subjects completed the full range of testing. Results showed that certain personality traits, like novelty seeking and a willingness to engage in actions that could lead to harm, were associated with a particular variant, RS1. Even better, men with longer versions of the RS3 showed greater activation of amygdala activity during the task, implying that the vasopressin-receptor gene affects how the amygdala responds as it processes social information.

Such a test doesn't say anything about actual behavior. The point is that people have different versions of *AVPR1A,* and that these different versions are associated with both personality and the function of the amygdala, a key part of the social brain. But when the same Israeli scientists who conducted the Reading the Mind in the Eyes experiment tried to match *AVPR1A* types with the way a group of just over 200 people performed in the dictator game, people with short versions of the RS3 variant gave significantly less money to another person than those with long versions did.

When the volunteers were surveyed as part of a standard examination of benevolence traits, people with long versions of the RS3 *AVPR1A* variant scored higher than those with short versions—as you might expect, since the short-RS3 people gave less money. Finally, the team examined the brains of human cadavers, finding that those with the long version of RS3 made much more of the receptor protein than those with the short version. The implication is that the more sensitive to vasopressin, the more likely one is to be sociable and engage in altruistic behavior.

As you recall, the prairie voles with a long version of *avpr1a* microsatellite were more attentive fathers to their young, and better bonders with their mates. Now consider this: Swedish researchers have a long-established program called the "twin offspring study." It's something like the venerable Framingham Heart Study (named for the Massachusetts town where most of the study subjects lived), which tracked the health of participants over decades. Swedish and American scientists, inspired by studies of pair-bonding in voles—and the fact that people, like voles, carry variants of the *AVPR1A* gene—decided to genotype hundreds of twin pairs and their spouses and partners, give them personality tests, and then survey the nature and quality of their relationships using a scale that measures pair-bonding behaviors among nonhuman primates.

Keep in mind that when scientists do these kinds of studies they almost never use the word "cause," as in, "Aha! Gene XYZ causes autism." Comparatively few diseases—cystic fibrosis, for example—are caused by one gene gone wrong. Human behavior is complicated, and so the favored term is often something like "link" or "association." (Versions of *AVPR1A* itself have been linked to anorexia nervosa and other eating disorders, as well as to "perfectionism" in children.) This careful wording reflects many uncertainties, the possible contributions of other factors, and the limitations of science.

Still, when the team sifted through all the gene data, tests, and surveys, it found a "significant" association between one variant of RS3 and male personality, behavior, and quality of relationships. What's more, this association was even stronger for men who were homozygous, meaning they'd inherited two copies of the variant. Women carrying it seemed unaffected.

According to the test of pair-bonding behaviors, men with one particular variant of RS3, called allele 334, scored significantly lower than men with none, and men who were homozygous scored lower still. As a group, those guys flunked Relationships 101. Naturally enough, when female partners were surveyed using an assessment that measures relationship cohesiveness, satisfaction, and affection, the men with one or two copies of 334 came out looking pretty bad.

Only 15 percent of the men who *didn't* carry 334 reported a marital crisis, including talk of divorce, during the previous year. In comparison, of the men who carried two copies of 334, 34 percent said they'd had a marital crisis and/or threat of divorce in the last year. Seventeen percent of the men who carried zero copies were unmarried, while 32 percent who were homozygous were unmarried. All the men in the sample were living with somebody at the time, and most couples had biological children together, but the men carrying this 334 version of RS3 were much more likely to be shacking up. Since bachelors who weren't living with a woman weren't even included in the sample, it's possible that an even greater percentage of that group would be homozygous for 334.

The particular version of the gene that was linked to all these traits is one of the most common in the human population.

While nobody can predict a man's personality based on one gene, these studies suggest that variations in this gene, which are linked to variations in the expression of vasopressin receptors and to activation of the amygdala, play a role in determining how the brain responds to social situations, thereby influencing behavior.

Environment and social experience also create variation. The male rats in Frances Champagne's experiments on maternal nurturing didn't show the same oxytocin-receptor effects from a mother's licking and grooming that the females did. Rather, male rodents show the epigenetic changes in vasopressin-receptor gene expression in their amygdalas. Male pups of high-LG mothers were more sensitive to vasopressin. Elliot Albers, of Georgia State University, has shown that socially isolated hamsters have different patterns of vasopressin sensitivity than hamsters living in groups. And the loners are more aggressive. Hamsters trained to fight react more to a vasopressin injection than those that aren't. Early hormonal exposures, which have organizing effects, can also tilt the way an individual's brain reacts to vasopressin.

Men are the same. We have brains that are organized to be more responsive to vasopressin than women's brains, which, in turn, helps drive our response to sexual cues, forming a love bond, and perceived threats to that bond. Mulcahy's story illustrates the power of these circuits to influence behavior, even at the price of betraying our own self-interest.

That our genetic inheritance, events inside our mother's uterus, and the way our parents nurture us—or don't—all have so much to do with a woman's love for a man, and with a man's love for a woman, can make the bonding paradigm seem awfully fragile, to say nothing about the effect of that whole pee-leads-to-monogamy thing. And Larry's belief that, brain-wise, a man is a woman's baby and a woman is an extension of a man's territory isn't the most politically cautious way to talk about human love. Many like to believe such notions are outdated stereotypes. They're not. We can fake it, but nature gets the last word.

But the bonding paradigm we've presented also leads to uncomfortable questions: Why, then, do people fall out of love? And, if bonding is so powerful, why do so many people who say they're monogamous wind up in bed with somebody other than the person they're supposedly being monogamous with?

7

ADDICTED TO LOVE

The bond between two people is powerful and certainly enthralling, but the nature of the glue that holds them together long past the early days of thrilling passion isn't exactly as usually portrayed in Victorian novels or in erectile-dysfunction-drug commercials. So, to learn more about the true nature of enduring monogamy, we've come to a man named Fred Murray. He's fifty-nine at the time of our conversation, mustachioed, short, with a round belly, but possessing a wound-up power that lends him an imaginary stature.

Murray has a lot to teach us about love because he was a drug addict, a stuff-the-glass-pipe-with-meth-and-crack addict. Drug addiction is a dark passage—Murray lived it for decades—that might seem as far removed from the joy of love as it's possible to be. But when Brian asks "Were you in love with drugs?" Murray smiles broadly and rolls his eyes at such a silly question.

Then he laughs. "Hell yeah, I loved drugs! I looovvved them," he says, turning "love" into a three-syllable word. This bit of phonetic magic doesn't satisfy him, though, because it fails to convey the strength of his ardor. He leans back in his chair and searches the ceiling for superlatives, but finds no stronger word for love than "love." "I mean, LOVED them. I loved them more than me. I loved drugs. Loved *them*. Loved getting them, having them, using them. Loved them. More than my wife. More than my daughter."

When Fred Murray says he was in love with drugs, he's not speaking in metaphors. He means it the way you might if you were to tell somebody how deeply in love you are with your spouse. He's right to say it this way, too, because the love he felt for drugs, and the love people feel for longtime partners, are the same thing. Love is an addiction. Some have even suggested that love is an addiction "disorder." We're not going that far, partly because love could well be an appropriate and healthy evolutionary adaptation, but love has the same hold on us, using the same brain mechanisms, as addictive drugs. The only difference is qualitative: drugs can be much more powerful than human bonds.

As we've hinted several times already, the brain processes activated during sexual bliss, and during the development of fetishes and partner preferences, have tremendous overlap with the brain circuitry that makes drug use feel so good. They both rely on most of the same structures, the same neurochemicals, and create the same changes in the brain. This can be true down to the level of individual cells. For example, when a rat takes methamphetamine, the drug stimulates some of the same neurons as sex does.

The parallels don't stop there, though. People who become regular drug abusers, and transition into addiction, soon find they enjoy the drug less. Likewise, the nature of love changes with time. We enjoy it less as the wild urgency recedes. Yet we remain in our relationships as initial passion turns into durable social monogamy because we've become addicted to each other.

The idea of love as addiction was floated back in the 1960s and then explored by Freudian psychoanalysts, though they mostly used the term "love addiction" to mean recurrent falling in love as a way to experience the excitement and pleasure of those initial romantic days of new love over and over again. But the way we are using the term is much older even than that. Plato likened love to "the tyrannical desire of drink."

The reward system we explored in chapter 3 is the seat of both drug

and love addiction. It evolved so we'd do what it takes to live and reproduce. Dopamine is a motivator. Without it, we wouldn't do much of anything. Mutant mice that don't make any are congenital couch potatoes
that won't move unless they feel pain or are very stressed. Humans with
Parkinson's disease, and whose dopamine is exhausted, remain almost
completely immobile. If not for dopamine, our ancient forebears would
certainly not have had the gumption to hike for miles in search of a wilde-
beest or to make more people by having sex.

Reward teaches us that eating a meal feels good enough that it's worth
hunting game or growing wheat. One thing humans quickly learned,
though, was that we didn't have to kill an antelope, or fight for access to a
sex partner, in order to reap pleasure. In fact, one could argue that the
search for more reward with less work is a central theme of the human
story, and over the millennia we've become very good at quickly, powerfully, and efficiently hijacking the circuits made to induce us into eating or
sex. The ancient Sumerians were brewing a form of beer at least as long as
five thousand years ago. Wine fermenting is similarly old. In the last century we've made Big Macs, bikinis, and that scourge of the latter-day bar
scene, Red Bull and vodka, all because they tickle reward circuits. Drugs
like cocaine, heroin, and methamphetamine are incredibly good at it.
They can be so seductive that users will overlook the well-advertised risk
of future pain in order to feel better now, sending themselves sliding into
a hole of addiction.

Murray, for example, no longer has a wife. He and his daughter have
rebuilt a relationship, and it's one he treasures, but he's plainspoken about
the years he didn't treasure it at all. Before he managed to stop drinking
and taking drugs, he'd lost a music career, a couple of jobs, his home, and
tried with increasing deliberateness to lose his life.

George Koob has spent decades trying to untangle the brain events
Murray and millions of others experience during the cycle of addiction.
Today, Koob, with the silver hair and mustache of an academic eminence,
is the chairman of the Committee on the Neurobiology of Addictive Disorders at the Scripps Research Institute in San Diego, California. He's

generally regarded as one of the world's experts on the brain and addiction. But that's not what he started his career doing. He began it in the 1970s trying to unravel the neurobiology of human emotion, especially the emotions surrounding stress and reward. He soon found the two fields—drug use and natural human emotion—were intimately linked.

Drugs are addictive, Koob tells us, because early on they make people feel wonderful with "a massive release, or a massive activation, of reward systems."

Details differ according to type of drug, but whether it's alcohol, heroin, cocaine, oxycodone, meth, they all work in essentially the same way. They prompt dopamine-producing neurons in the mesolimbic reward system to pump out a flood of the neurochemical. Stimulants like cocaine do it directly and quickly, while alcohol does it somewhat less directly and more slowly. Whatever the route, the nucleus accumbens acts as a reward central switching station, sending the resulting signals to the amygdala which assigns a salience—usually something like "nice!"—and to the ventral pallidum (the area Larry and his lab colleagues found packed with vasopressin receptors in monogamous voles). The amygdala also forwards the information to other brain structures like the bed nucleus of the stria terminalis (BNST), one of the regions where Dick Swaab found sexual orientation differences.

Animals treated with drugs of abuse, and exposed to a stimulus, quickly learn to associate the stimulus with the good feeling. After just a few drug sessions, they begin to receive a burst of dopamine just by anticipating the stimulus even if the drugs are nowhere around, like the rats who found the light on the wall pleasing after learning to associate it with sex. The learning works on people, too. Imaging studies have shown that in response to cues like a crack pipe, the brain of an addict will respond in much the same way as if he or she had access to crack itself. When the circuit senses the drug-related cue, its connections with the motor system trigger directed motivation toward a goal—seeking a drug buy, snorting a line, slamming down a shot of tequila. Fourteen-year-old heterosexual boys experience exactly the same motivational process with the arrival of the *Sports Illustrated* swimsuit issue in the mailbox or by clicking on an Internet link leading to Japanese cosplay porn.

Generally, this drive is associated with positive feelings. A drug user can be filled with pleasant anticipation about the next time they'll use. If they make plans for doing so, perhaps engaging in a drug-related ceremony like preparing a crack pipe, dicing up cocaine with a razor blade on a special mirror, or retreating to a special place to get high, and then consummate those plans, they enjoy drugs even more. If they're surprised with, say, a few lines of cocaine without expecting to encounter cocaine, they'll be happy to snort it, but it won't be as intense an experience as if they'd spent more time looking forward to the moment. Once that moment arrives, the reward system fires even before the drugs are taken. Then, with the drug, the brain is filled with hedonic sensations. After, users feel the calming rush of brain opioids.

The same mechanism explains why foreplay works to heighten erotic joy. Just as drug users like drugs more if they anticipate using them, and engage in ceremony before taking them, the tease and delay before the consummation of sex sensitizes the reward system.

Over time, when the cues we've learned to associate with good feelings are present, our prefrontal cortex is muted and it's pretty easy to ignore everything else to focus instead on salving our appetite. We can become so highly motivated, we can act impulsively. If you happen to be a drug user, you might excuse yourself from a lunch meeting to snort a line of cocaine. If the cues are erotic, you could wind up in flagrante delicto on the hood of a 2003 Ford Focus.

While that kind of giddiness can be fun, "unfortunately," Koob says, "the cycle can't be sustained."

Eventually, drug users become like the singer of Cole Porter's classic song. They can't get a "kick from cocaine," or at least not the same kick from the same amount, and "mere alcohol" wouldn't thrill them at all. So they have to use more, and then still more, to achieve the same effect they once felt from a small amount.

You've probably heard, or read about, or seen some version of this part of the drug addiction story—the tolerance to the drug and the escalating use. It's been the denouement of dozens of rock band breakup exposés. And depictions of the physical signs of addiction and withdrawal have been around since ancient times. "But that is not the end of the story," Koob continues, "though most people end there." The most relevant part for our discussion of long-term monogamy is what comes next: the reason why rock singers and celebrities and Fred Murray, even seeing the damage, cannot, or will not, stay off drugs.

NUMBER 1 WITH A BULLET

Murray was born and raised in Gary, Indiana, in the shadow of the giant U.S. Steel works. Even in its heyday, Gary could be a tough town. Murray's father was a welder. His mother worked in U.S. Steel's food services department. Both drank, a lot. Murray remembers the first time he got drunk. He was six years old and he fell down the stairs. His father left home when Murray was still a young boy, and his mother's drinking got worse. He moved in with his grandmother, but Murray spent most of his time in the streets of Gary. By his junior year in high school, he believes, he was an alcoholic. "I needed to drink," he recalls. "I told myself I could not drink, could stop any time I wanted to, but that was a lie."

Junior year also marked his first arrest. His little brother had found a wallet with money inside, and two older, bigger boys took it from him. Murray decided to get it back. He pocketed a small-caliber pistol he kept for protection, found the boys, and threatened to shoot them if they didn't return the wallet. They did, but then, Murray claims, they went to the police and accused him of holding them up on this occasion, and on others. Murray got one to five years in the state reformatory. Older inmates there taught him how to strain shoe dye through a piece of bread to get at the alcohol. Sometimes, while picking up trash around Indiana's Raccoon Lake State Park during work details, he'd find half-empty beer cans or wine bottles and drink the leavings.

A judge let Murray out early, after only two months. He'd caught a break and he knew it, but the first thing he did after his release was buy booze. "Drinking was my main thing," he recalls. "Crown Royal, man, that was my favorite. But if I couldn't afford it, I'd drink beer."

The booze didn't seem to get in his way, though, mainly, he thinks, because he'd become tolerant. He graduated from high school, began working for a railroad, became a foreman, got married, fathered a little girl. He'd long been interested in music. Gary had a tradition of nurturing rhythm-and-blues groups, such as the Spaniels, who had a big hit in the fifties with "Goodnight Sweetheart Goodnight," and the Jackson 5, and Murray, who used to peep into the Jacksons' practice sessions, began singing a little himself, and writing lyrics. He helped form an R&B band of his own called The Group. The band competed with other Gary- and Chicago-area outfits for the opening slot when big names came to the region. The Group opened for Gladys Knight, Ray Charles, Earth Wind & Fire. There was talk of touring and a recording contract.

One night, when Murray was twenty-seven, the lead singer for a famous national act and some recording-industry types attended a performance by The Group to scout the band. "My manager and some big shots were in this room doing lines of cocaine," he recalls. "I had never seen lines that big, that thick." One of the men demanded that Murray snort some as a way to guarantee that Murray wouldn't inform anybody about the drug use.

"Then I got up onstage, man, and it was like I thought I was the coolest, best—I was the worst singer in the group—but onstage, I was all over that stage! It was like I had something extra. The audience was going crazy. I remember telling my manager, 'Hey, that stuff made me OK. I gotta get some more.'"

Murray and other band members started using cocaine with increasing frequency. "It seemed like it gave you that sense of energy, excitement. And talk about social lubricant! Man, it's almost like you get a straight adrenaline shot to the ego, and all the sudden women I'd never say a word to, I'm talking to them."

He looked forward to snorting coke. If he couldn't get it, he'd drink—alcohol was always his reliable standby—but cocaine was a lot more pleasing.

Murray became very impulsive, just as Koob predicts. One day he decided to quit his railroad job.

"I walked into the personnel office and this guy, Leroy, I will never forget him. He said 'Fred, do me a favor. Go to the bathroom and look at your face.' I went into the bathroom and both my nostrils were white with powder. I cleaned myself up, came back out, and Leroy said, 'What the hell is wrong with you? What are you doing?' And I said, 'I'm retiring and I want the pension money.' He said, 'Fred, leave it. Do not touch it.' I said, 'Leroy, I want it all.' And he said, 'If you take it it'll be gone in six months.' Well, he talked me out of it, and as a result, I still have some pension today."

Leroy was doing the job Murray's prefrontal cortex should have been doing but couldn't, because Murray's brain was so chronically full of dopamine.

As we explained in chapter 3, the same effect happens, to a much lesser degree, in an acute way in all of us. For example, in 2010, a mixed-disciplinary group of economics and psychology researchers in the United Kingdom conducted tests on people who were given the drug L-dopa, a dopamine replacement often used by Parkinson's disease patients. The experiment used a series of tasks demanding intertemporal choice, a device that gained fame when Walter Mischel used it in the late 1960s to study "delayed gratification" in children. Such experiments usually demand subjects choose from two different options, separated by time, like presenting children with a piece of candy and telling them they can eat the one piece now or wait fifteen minutes and receive three pieces later. The UK L-dopa studies used money. Subjects could have a small sum of money right away or a significantly larger amount many weeks later. The most economically rational choice would be to delay gratification and earn the bigger payout—the choice Leroy urged Murray to make. But after being given L-dopa, the test subjects chose the smaller-sooner option much more often than when these very same people were given a placebo. With dopamine, they discounted the value of the future reward, which made the sooner reward relatively more attractive than it should have been if they'd been dispassionately weighing their options.

When Belgian researchers used intertemporal choice to test 358 young men, they didn't have to use L-dopa. They simply exposed the men to pictures of pretty women in bikinis or lingerie, video of women in bikinis running over "hills, fields, and beaches" *Baywatch*-style, women's bras, and neutral imagery of lovely landscapes. The men were told to negotiate a payout: fifteen Euros now or a larger sum later. As a group, the men exposed to the sexy women negotiated for smaller payouts sooner than the men who'd seen the landscapes.

Fred Murray often thought he was in control and making perfectly rational choices. When he'd slip into a liquor store, pick up a bottle of Crown Royal, unscrew the cap, take a couple of big swigs, and then put the bottle back, it made sense. "It wasn't stealing," he recalls, "because I didn't walk out with the bottle." It made sense when he broke into a car.

With time, though, this exciting impulsivity faded into paranoia. His increasing use led to dealing. Despite trying to keep his drug life separate from his home life and work life, he began to feel he was looking through the wrong end of a telescope that narrowed his field of vision to drugs— how to get them, sell them, take them. After he began smoking crack, he named his pipe Sherlock because it was shaped like the celluloid Holmes's antique calabash. He treated Sherlock like "my best friend. Every time after I used it, I'd wrap it up in this soft towel, place it in a special drawer. It was important to me."

Fred Murray was now in love with drugs. "If it told me to stop hanging around certain people, I'd stop," he says, lending drugs the status of personhood. "If it said 'Walk in that store, pick up that VCR, and walk out,' I would do it. 'Call your job right now! Tell them you ain't coming. Right now!' and I'd say 'OK.'"

The problem, as Koob explains, is that an addict's brain changes. "It adapts to the drugs. The reward system can be compromised."

Chronic drug abuse reconfigures the mesolimbic dopamine system,

194 • THE CHEMISTRY BETWEEN US

especially how dopamine acts in the nucleus accumbens. The transformation is like a switch, changing the desire for drugs from "liking" to "wanting." At first, dopamine action provides motivation that kick-starts an addict's pleasant appetite. It stamps the brain with positive associations of the contextual cues that enhance eagerness to consume. Later, though, the addict becomes subject to what Koob calls "negative motivation." Rather than keen expectancy, euphoria, and impulsiveness, the drive now is full of anxiety, dysphoria, and compulsion. The addict feels compelled to act because if he doesn't, something bad is going to happen.

"What you are worried about is feeling terrible when you're not on the drug," Koob says. "That's the dark side, the negative reinforcement. It's insidious." For one thing, this compulsion doesn't merely turn down the volume of the prefrontal cortex, it well nigh disables it. Imagine, Koob suggests, what our lives would be like without it. "You would always seek immediate rewards rather than delayed rewards, and you'll always be seeking something to assuage the immediate problem rather than working through unpleasantness." That's the life of an addict. That life is not about getting high and feeling good, it's about trying to avoid feeling bad.

Murray just says, "It was like I was a slave. I mean, I *was* a slave."

Other, natural, sources of reward now make no impression. Addicts lose interest in family, work, even eating. Murray stopped having sex. "What can't get up, can't get out," he recalls. Drugs, he says, rendered him not only uninterested but impotent. "Noodles for freakin' days, man. Once I started smoking crack, it was, like, 'Cancel Christmas, baby!'"

The primary weapon of negative motivation is a brain hormone called corticotropin-releasing factor, or CRF. It engages the body's hypothalamic/pituitary/adrenal (HPA) axis. "Your amygdala and CRF system go nuts," Koob explains. "That's the fight-or-flight response. So you get a double whammy with drug addiction. You lose the reward, but you activate the brain stress system."

Imagine being chased up a tree by a bear, Koob suggests. You see the bear running toward you, and CRF rushes into neural receptors, which in turn triggers the HPA axis. You feel a sudden burst of energy giving you the power to run for your life and climb a tree. Once you're up the tree, and you realize the bear can't climb it, opioids are released from the ventral tegmental area (VTA). You calm down and are able to think about how to get yourself out of this predicament. With luck, the bear leaves, you climb back to the ground, then start rehearsing the story you're going to tell over a glass of wine back at the lodge.

But if the opioid-producing VTA never kicked in, you'd stay up that tree long after the bear leaves, too terrified to come down. This is what happens to an addict. He can wreck the natural feedback so CRF and the HPA axis just keep pounding away, demanding action to stop feeling so panicky and miserably stressed. Since natural reward is no longer effective at relieving stress, the only action that'll work is to take more drugs. Just as you wouldn't stop sprinting toward the tree while being chased by the bear to consider whether you'll order red or white wine later, for a drug addict under this stress, all other concerns—loved ones, jobs, hobbies, even aversions to crime and risk—fade away.

Murray was trapped. Not being able to stop filled him with such self-loathing, he planned suicide—twice. Finally, in 1994, after another brush with the law related to dealing, he removed Sherlock from its special place and smashed it on the ground. "I felt like I lost my best friend," he recalls. "I mourned." He packed a few things and moved to Oceanside, California, where a family member lived, and got a job making clubs at Callaway Golf. It seemed like a clean break, but his craving for cocaine returned when he overheard coworkers talking about a drug deal.

Any cue can trip the CRF system, even long after an addict has recovered from the physical signs of withdrawal. Many longtime smokers experience this when they attend a party or walk by an office building entrance where people are standing around puffing on cigarettes. This, Koob explains, is why so many addicts relapse. The reactivated stress response compels them to use, even knowing they no longer like the drug, that it damaged them, and, in many cases, led to serious personal loss.

Within months, Murray was snorting powder, smoking crack, and now using crystal meth, too. "I lost my apartment. I completely stopped talking to my sister. I wound up sleeping in a closet in an empty apartment even though I had a job because my disease told me that if I spent anything for shelter, it would punish me. So all my money—all my money— after my eviction—every penny—went to drugs. I'd get enough drugs so I could sit in that closet and get high all day then go to work at night. I don't even remember if I slept."

VOLES ON DRUGS

Scientists like Koob have learned so much about drug addiction partly because they've been able to turn all kinds of animals into drug addicts. Drugs from alcohol to meth work in rats, mice, and monkeys in much the same way as they work in people, down to the receptor level. So does the arc of addiction. A cokehead rat will constantly seek out cocaine, for example, and will, if left alone with a big stash, kill himself with it. He'll attach great meaning to a lever he pushed to get cocaine even if the lever no longer delivers it. He's become a lever fetishist because the lever itself triggers brain reward. If his drugs are taken away, he'll go through withdrawal, and then, later, after the physical symptoms have disappeared, in response to a drug-related cue, he will feel compelled to find cocaine because his CRF system will be activating the HPA axis making him a very stressed-out rat.

Prairie voles have recently proven to be valuable in addiction research, too. Thanks to their ability to bond with partners, they can model how drugs affect social relationships, and, in the process, further illuminate the nature of the pair-bond.

In 2010 and 2011, Zuoxin Wang's lab at Florida State conducted a series of experiments to explore the intersection of drugs and the vole version of love. When they gave virgin male voles doses of amphetamines, the voles formed a place preference for the chamber in which they received the drug, just as experiments have shown other rodents, and people, do when they're given amphetamines. Remember that prairie vole virgins also form a partner preference with mating. Yet, when they gave the drug to virgin male voles, and then allowed them to mate, the voles on drugs did *not* form a partner preference unlike almost all other prairie voles. Also, males given the amphetamine *after* forming a partner bond didn't show much interest at all in the drug. They didn't form a place preference. It was as if prairie voles have one chance to bond to something, and once bonded to that thing—drug or another vole—there's no room to bond with anything else.

This is just about what happens. Studies in humans and animals have shown that drug addiction can place the nucleus accumbens into a kind of deep freeze, blunting neuroplasticity. The reward system loses much of its ability to react to new, potentially lovely stimuli like, for example, one's first taste of truffles, one's new baby, or a new lover. In the Wang lab's experiments, pair-bonding took most of the fun out of drugs, and drugs took most of the fun out of bonding.

One of Wang's students, Brandon Aragona, deciphered a key to this phenomenon. Using male voles, he discovered that the voles' brain-reward system makes the same kind of switch seen in drug addicts. The system has to enable virgin prairie voles to find the idea of mating, and then bonding, attractive. It has to motivate them to do it. If it didn't, there'd be no more prairie voles. After one day of mating, the voles will have formed a partner preference for each other but not yet be firmly bonded. In other words, the system will still be turned on in its original configuration. The male, for instance, will still be willing to entertain a strange female who happens to wander by.

If the system stayed like this, there'd never be a monogamous bond because the voles would be equally interested in mating, then forming new bonds with different partners, something like their meadow vole cousins. Somehow, the system has to be turned away from mating and bonding with new partners and toward maintaining the bond that's now formed.

The secret is time. Twenty-four hours is too short. But within a matter of days, the male vole has cemented his bond with his mate, and he will now brutally attack an intruder female. Aragona's experiments showed that the reward system itself undergoes a change with bonding. The vole nucleus accumbens is reorganized, rendered less plastic just as Koob describes it happening in a drug addict. In effect, the dopamine system first makes a prospective sex partner look good, then, later, it changes so the voles will narrow their world to stay with that partner. A few of the pairbonded voles, 28 percent, did not undergo the required reorganization, and interestingly, in the wild, about 20 percent of previously bonded prairie voles will indeed form a second bond with a new partner.

This switch in the accumbens also addresses another possible problem common to human relationships: boredom. The voles could simply lose interest in each other and wander off. Yet when either one of the two partners leaves the nest to forage for food, they return time after time. Like people, birds, and Dorothy in *The Wizard of Oz*, they feel compelled to go home. What makes them come back?

Larry has long believed that experiencing a loss, such as separation from a partner or a partner's death, is akin to an addict without drugs, that negative feelings accompanying loss motivate individuals to maintain bonds. A German scientist named Oliver Bosch, who normally specializes in the study of maternal behavior, came to Larry's lab to conduct experiments to test this idea on the voles. By the time he and Larry had finished, they'd uncovered an important mechanism of monogamy.

Though he typically uses rats and mice in his research, Bosch was intrigued by the pair-bonding of the voles. "In our lab in Germany, we saw some outcomes of separating mothers and offspring," he recalls. "And voles have this different type of adult bonding. So we wanted to see what happens if you interrupt that."

Bosch, a gregarious man with short brown hair and oval spectacles, works at Regensburg University about an hour outside Munich in his home state of Bavaria. The modern campus, built in the 1960s with the techno-optimism design sensibility so common of the era, seems at odds with the old city it adjoins, and a little at odds with Bosch's own sensibilities. Though he speaks with typical scientific caution about translating rodent behavior into that of people, he's not studying rats because he's worried about rats; he's worried about people, especially modern society's effects on all of us. He believes social engagement, especially parent-child interaction, is a key to happiness.

Meeting Bosch has the effect of making you want to run home to your mother. "I was at a conference recently, and this guy from Australia was there," Bosch begins a typical story. "It was his birthday and he was far away from home. He told me he got a birthday cake from his wife and his really young kid. But he was not able to get a hug! This is what caused him to be sad, not touching them, and a hug is very important!" The good thing about Germany, he says, isn't the trains or Porsches or skiing. It's that most people still live somewhat near family, and so "to get a hug in Germany is still possible."

To investigate the rodent version of getting hugs, and what happens in the absence of hugs from a bonded partner, Bosch took virgin males and set them up in vole apartments with roommates—either a brother they hadn't seen in a long time or an unfamiliar virgin female. As males and females are wont to do, the boy-girl roommates mated and formed a bond. After five days, he split up half the brother pairs, and half the male-female pairs, creating what amounted to involuntary vole divorce. Then he put the voles through a series of behavioral tests.

The first is called the forced-swim test. Bosch likens it to an old Bavarian proverb about two mice who fall into a bucket of milk. One mouse does nothing and drowns. The other tries to swim so furiously the milk turns into butter and the mouse escapes. Paddling is typically what rodents will do if they find themselves in water; they'll swim like crazy because they think they'll drown if they don't. (Actually, they'll float but apparently no rodent floaters have ever returned to fill in the rest of the tribe.)

The voles that were separated from their brothers paddled manically. So did the voles who stayed with their brothers and the voles who stayed with their female mates. Only the males who'd gone through vole divorce floated listlessly as if they didn't care whether they drowned.

"It was amazing," Bosch recalls. "For minutes, they would just float.

You can watch the video and without knowing which group they were in, you can easily tell if it's an animal separated from their partner, or still with their partner." Watching the videos of them bob limply, it's easy to imagine them moaning out "Ain't No Sunshine When She's Gone" with their tiny vole voices.

Next Bosch subjected the voles to a tail-suspension test. This test uses the highly sophisticated technique of duct taping the end of an animal's tail to a stick and suspending it. As in the swim test, a rodent thus suspended will usually flail and spin his legs like a cartoon character who's run off the edge of a cliff. Once again, though, while the other males did just that, the divorced males hung like wet laundry.

In a final behavior test, Bosch placed the voles on an elevated maze, like the ones we've already described that tested anxiety. On such a maze, the animal's desire to investigate fights with its fear of exposed areas. Compared to the other voles, the divorced males were significantly less likely to explore the open arms of the maze.

All these tests, commonly used to test lab animals for depression, showed that if you separate a pair-bonded male vole from his mate, you'll get a very mopey vole who uses what's called passive-stress coping to deal with the overwhelming anxiety of partner loss. "When the separation takes place, this is what causes the animals to feel so bad," Bosch explains. "We found this increased depressive behavior and that tells us the animal is not feeling well." He doesn't mean "under the weather," he means the divorced voles are emotionally miserable. "It is like when my wife went to the States for a post-doc for one year, so I knew I wouldn't see her for at least six months. Well, I was sitting at home, laying on the couch, not motivated to do anything, not to go out and meet friends like I usually would."

Koob and others have used drugs to create the very same behavior in other lab animals. When the drugs are taken away from rats and mice, they display the same passive responses to elevated mazes. They withdraw socially. They mope. Human addicts do the same, Koob points out, mentioning characters in movies like *Leaving Las Vegas* and *Trainspotting* as examples.

To explain the physiology behind this passive depression state in the separated voles, Bosch checked their chemistry. The males separated from their mates had much higher levels of corticosterone, a stress chemical, in their blood than did any of the other groups, including voles separated from their brothers. Their HPA axis was working so hard, their adrenal glands weighed more. Bosch nailed CRF's role in driving both the HPA axis overdrive and the mopey behavior by blocking CRF receptors in the voles' brains. When he did, the divorced voles no longer hung limply from the sticks. They didn't float for as long in the water. They still remembered their mates, and were still bonded to them; they just didn't worry about it when they left them.

But here's the strange thing: both the voles who stayed with their female mates and the voles who were forced to split from the females had much more CRF in the BNST than did males who lived with, or were separated from, their brothers. In other words, loads of this stress-related hormone were being pumped in *both* the voles who got depressed after separation and voles who were still happily bonded and didn't show signs of passive-stress coping.

"Bonding itself produces high CRF," Bosch says. "But this does not mean the system is also firing." There is something fundamental about living with a mate that results in more CRF stress hormone in the brain, but that also prevents the engagement of the HPA stress axis as long as the mates stay together. Using an interesting metaphor for bonding, Bosch says "I compare it to a rifle. As soon as they form a pair-bond, the rifle is loaded with a bullet. But the trigger isn't pulled unless there is separation." He thinks that vasopressin serves as the chemical trigger to fire off the HPA axis during separation, though the exact roles of both oxytocin and vasopressin are still unclear.

Addicted drug users load the rifle, too. The gun won't fire unless they stop taking the drug. For the bonded voles, "it won't fire unless the partner leaves the nest," Bosch says. "This pre-loading allows the system to work very fast. As soon as the separation takes place, this is what causes the animals to feel so bad." That bad feeling drives the voles home. "You want to make sure this feeling goes away and the only thing for the animals to do is go back to the partner." Once back home, oxytocin may be involved in helping ease the anxiety the separation caused. The rifle stops firing, and the stress system goes back to normal.

For humans, falling in love is like putting a gun to your head. You're lured into a relationship, enjoy its pleasures, and then, over time, those pleasures fade and compulsion takes over. "It is a quite similar situation, you know, to when a person at first has the high feeling in a relationship, the CRF is silent, and the dopamine reward is taking over," Bosch says. "You feel high. Everything is cool. Everything is nice. Then, after a while, nature makes sure you still want to stay with a partner. And this system makes you feel sick as soon as you leave the mate. This is the idea of the whole story."

We ask if he believes this means the voles are returning because they are still positively motivated to be with their partners—the "liking" phase—or because they want the misery of the separation to stop—the "wanting" phase. They want the misery to stop, he answers. "We have this normal together, whatever that normal is. And the bad feeling forces you to come back."

Koob agrees. The CRF system is there, he explains, to signal that a loss has occurred and we need to do something about it. Likewise, when rats are taken off drugs, and their brains tested in real time for CRF levels, he finds a large boost in CRF in the relevant reward regions. When alcoholic rats denied booze are given the same CRF-blocking drug Bosch used in the voles, they stop drinking excessively even if given access to alcohol, and they don't show the same passive-stress coping.

Parenting follows the same pattern, further supporting the idea that love between adults has its roots in parent-baby bonding. As we've explained, nurturing is rewarding. If it weren't, we wouldn't do it, just like we wouldn't have sex or fall in love. And like love, it shares pathways with drug addiction, including, among others, the amygdala, the VTA, and the nucleus accumbens. Parents "fall in love" with their babies, but in time, as with a love match between adults, there's the risk of boredom, not to mention aversion. After a number of sleepless nights, dirty-diaper changes, and general baby crankiness, the bliss of the early days can turn into drudgery. Parental loss of interest is a life-and-death matter for a baby, so nature built in a system to make sure parents feel compelled to give care whether or not they like doing it.

When a mother loses her child in a shopping mall, her CRF level spikes. When she finds the child, opioids calm her. When a baby cries in distress, CRF activates the HPA axis, driving a parent to pay attention to the child, not so much because caring and nurturing are as pleasing as they might have been during the days after the baby's birth, but because the parent is now negatively motivated to make the stress stop. With nurturing and contact, oxytocin release turns down the HPA thermostat, and feelings return to normal.

This could help explain why drug-abusing mothers often neglect their babies. Studies of new mothers on cocaine show that they're less engaged and responsive, likely because drugs blunt the salience of natural rewards, making a parent less eager to nurture, just as drug addiction makes addicts less engaged in social relationships of all kinds, and giving amphetamines to voles stops them from forming a partner preference. Drugs have interfered with their natural ability to bond with their children.

Bosch's results were in males. They generally track with studies of females, but there are some important differences. Female voles given amphetamines react like males, only more so. They seem more vulnerable to the rewards of the drugs and the place preferences that drug taking induces. In Bosch's experiments, males separated from their male buddies didn't grieve the separation. But females separated from other females, like cage mates and sisters they've lived with for a long time—any female with which they've shared social support—do grieve. Males put all their emotional capital in one bank, with their female partner. Females will display depression behavior if they lose their mothers, sisters, close female friends, or their bonded mates, perhaps a clue that helps explain why women suffer from depression at roughly double the rate as men.

DRUNK DIALING EXPLAINED

"Drug addiction and love are absolutely parallel," Koob declares without a hint of doubt. That sure explains a lot of crazy behavior.

Think of love's arc. Two people, strangers to each other, with dreams, goals, and paths of their own, meet. There's an attraction, an affiliation, sex. Seemingly in no time at all, their thoughts become what in almost any other realm of life would be considered a form of obsession, very like the narrow focus of a drug addict. The smell of the back of her neck, the texture of the hair on his chest, the softness of her lips, the sound of his voice when he whispers urgent erotic pleadings in her ear, the Lautrec prints on his walls, her collection of *Vogue* magazines—all these sensory cues are not only sharply vivid but also important for inexplicable reasons. The mere thought of her perfume can distract him from his work for long minutes of reverie. Then one day life plans are changed because the consequences of not changing them are too painful to consider. Years pass. He wonders why she insists on keeping those damn magazines. She finds him pedantically boring, and by the way, Lautrec prints are a decorating cliché. Yet they say they are happy. Not happy in the way they once were, but stable, safe, comfortable. When she travels for work, she misses him, and their home, and he misses her. Life didn't turn out the way either of them dreamed, but whose does? They loaded the gun and now they live with it pointed at their heads.

This is not necessarily as cynical a view as it might seem. Having a gun pointed at you could help keep you on a path that yields the most happiness over the longest term, not to mention that it helps satisfy the evolutionary mandate to raise children.

There's a lot of indirect evidence to support the love-as-addiction scenario. Lovers, for example, can act like painkilling drugs for each other. In one test, fifteen people who were nine months into new relationships—long enough to be in love, but not so long as to have grown sick of each other—were placed in fMRI machines and subjected to varying degrees of pain caused by heat. They looked at pictures of attractive acquaintances, their partners, and words that were previously shown to reduce pain. The words did reduce the amount of pain the people said they were feeling, but imaging revealed that the reduction was due to simple distraction: the words took their minds off it. Pictures of acquaintances had no effect on pain. But pictures of their lovers reduced pain by activating the reward system, including the accumbens, amygdala, and prefrontal cortex, just like drugs of abuse do.

The addiction model also explains the allure of the long-distance relationship. It's really just extended foreplay. When appetitive reward is won only intermittently, explains Jim Pfaus, "not only do you not get tolerant, you are actually more sensitized to it. Sex works the same way, and that's how long-distance relationships work. It's dramatic, right? 'We'll see each other every two weeks.' And then you anticipate, and for two days before you see each other, you are full of anticipation for the reward, and then engaging in that appetitive response, as well as hot sex, well, that glues those cortical representations of your partner." Neither of you ever has a chance to become bored. You're both forever fascinating to each other, a relationship idling in that happy intermediate zone between first irresistible passion, and later disgust over the wearing of sweatpants to bed. "If you can get it every night," Pfaus explains, "it's like masturbation. The degree of reward goes down. Even the ejaculation decreases in size and volume! People who study drug addiction know this; if you give cocaine every day, you become tolerant of the cocaine."

Addiction certainly helps explain our behavior when love goes bad. Like Murray's destruction of Sherlock, the end is traumatic no matter who pulls the trigger. "Drug addiction is very similar to love relationships breaking up," Koob says, thinking of Bosch's experiment. "I think this is what the system is meant to do, to bring you back to your partner. It's there for a purpose, to bring you home to be with your mate."

The voles separated from their bonded partners could be said to be grieving, and people grieve, too, of course, both the loss of love and the death of a loved one. Mary-Frances O'Connor, a neuroscientist at the University of California at Los Angeles, examined the brains of women who

206 • THE CHEMISTRY BETWEEN US

had recently experienced the death of a sister or mother. Some of the women were suffering what's called "complicated grief." More than sadness, complicated grief is a chronic, pathological yearning for, and preoccupation with, the deceased. Other women felt "non-complicated grief." While the women were in an fMRI, O'Connor exposed them to pictures of their dead loved ones, or pictures of strangers, each picture paired with words related to grief or with neutral words.

The women with complicated grief displayed more intense activation of reward structures. That may seem odd unless their grief is seen in light of addiction. When the words reflected grief, only the women with complicated grief showed activation of the nucleus accumbens. This accumbens activation was associated with more yearning for the dead person very much like the wanting of a drug, or the way the smell of on old T-shirt can bring back a rush of painful yearning for a lover who's left.

The immediate stress of a breakup, and the chronic stress after, can be so great it can affect our health. When people go through a marital separation, they show significant declines in immune system strength. Newly separated people visit doctors more often, have more acute and chronic health problems than married people, die from infections at a greater rate. Contrary to the usual image of a newly separated man celebrating his "freedom" by prowling nightclubs for beddable women, men appear to suffer more severely, perhaps because men invest all their emotional bonding in their partner, whereas women often enjoy emotional and social support from other women.

Far from being able to concentrate on anything as trivial as work, the newly single can find themselves pathetically, and obsessively, searching for any sensation connected to the departed loved one: a bit of hair, a scrap of paper with handwriting, the flavor of a favorite food. People will spend long minutes staring at photographs. Likewise, drug addicts who've stopped using still maintain heightened and preferential attention to any drug-related cue. The power of these cues emanates from their ability to reactivate CRF and the HPA axis, creating a compulsive craving for contact—the reason why Murray destroyed Sherlock. Even without

Sherlock, the slightest hint could lead to temptation for years after he stopped using. "If I walked into a kitchen and saw Coffee-Mate powder spilled on the counter? I tell you, man, I'd have to wipe it off right away or even leave the room."

Under the thumb of the stress system, we will do things we never imagined doing, like drunk dialing the departed lover at 2:00 a.m. or listening to sad Edith Piaf songs though we haven't any idea what the words mean. We drink, especially men, who release much more dopamine with alcohol than women do. We are out of control because we're denied access to the one natural source of relief—the person we love.

In animals, CRF induces drug relapse, and in humans, the stress of a breakup leaves both former partners vulnerable to breakup sex. Dumpers may rationalize by thinking that perhaps they made a big mistake by ending the relationship. Dumpees aren't thinking of self-respect because they're rationalizing with "Aha! She wants me back!"

Murray rationalized all the time. "I'd say 'Yeah, it's OK to not pay Montgomery Ward. I'd tell myself I'd double up the payment next time. Or, 'It's OK to not go pick up your daughter. Somebody else will do it.' Or, 'It's OK to be late. It's OK to not go at all. You're busy, it's OK for you to not go to band practice, you're too good anyway.'"

Love-as-addiction is often responsible for the rebound relationship. If we were all voles, Koob says, and had just broken up with our mates, we'd do better by not pouting and taking action instead. "The best thing to do, of course, would be to wait for another vole, or just go find another vole, to be the Newt Gingrich of voles." As we've said, sex itself is not an addiction, contrary to what some pop psychologists and "sex rehab" entrepreneurs say. But sex does trigger oxytocin release, which quiets the heightened stress response to separation from the loved one. People who are newly separated from a love relationship, even if they initiated the breakup, can still experience the drive to relieve that stress, leading them into a new pairing.

A few people respond to lost love with extreme behaviors, becoming stalkers or committing suicide. One review of notes left by Americans who committed suicide showed that love was a more common motive than lack of achievement for both men and women. In India, about ten people every day kill themselves over lost love—more than poverty, unemployment, bankruptcy. Interestingly, depression is another significant factor in suicides, and people with depression often have chronically high CRF levels because the stress response is stuck in the "on" position.

Merely contemplating a breakup can create the stress and fear of the real thing. When college freshmen in love were asked to predict how upset they'd be, and for how long, if their partner ended the relationship, those who described themselves as more in love, unlikely to begin a new romance, and didn't want the breakup significantly overestimated how bad they would feel and for how long. This is undoubtedly why some people choose to remain in a relationship after some transgression by one of the partners.

When Silda Spitzer, wife of then New York governor Eliot Spitzer, chose to appear with her husband in front of news cameras, and to remain in the marriage, after he was accused of visiting a prostitute, she was denigrated by some for being an antifeminist doormat. She even inspired a TV series, *The Good Wife*. Yet many husbands and wives stay in relationships despite the sexual infidelity of a partner. No doubt religious belief, economics, and children all play roles in these decisions, but addiction is a powerful visceral motivator.

Even men and women who have been verbally or physically abused by their partners refuse to leave those relationships, just as Murray couldn't leave his relationship with drugs. They rationalize their choice to stay, for example, by focusing on positive traits their partner might possess. Those who remain, but then do manage to leave later, often look back on their former mental state as being "brainwashed" or confused. When Murray uses phrases like "My disease told me . . . ," he's reflecting the same state of mind.

Like the 28 percent of voles that did not show the accumbens reorganization in Aragona's experiments, our willingness to love, and our eagerness to bond, the degree to which any of us will be vulnerable to monogamy addiction, or to extreme manifestations of that addiction, depend upon the same kinds of genetic variations and environmental circumstances that influence a tendency toward drug addiction. For example, positive social interactions make us feel good partly because they result in release of those brain opioids. When scientists examined over two hundred people for variation in the mu-opioid receptor gene, they found that those carrying a certain polymorphism were more likely to become romantically entangled, and received more pleasure from doing so, than those carrying another polymorphism. This version of the gene has also been associated with increased drug highs and a heightened stress response compared to other polymorphisms.

Fred Murray believes that he was almost destined to become a drug abuser given his family history of alcoholism. He may be right, or it may be that his environment, including rough streets and lousy parenting, was more influential. Likely, it was a combination. But he was most definitely in love with drugs and felt all the pain of trying to break up.

In the end, he forced a climax. He checked himself into a cheap California motel room and began smoking a combination of crack and meth, as much as he could. As he'd hoped, he could feel his heart pounding in his chest. Sweat poured out of his body. Just a couple more bowls and his heart would burst. But he passed out. When he woke, he looked at himself in the mirror and was overcome with disgust. "I couldn't even do that right," he says of trying to kill himself. He called Callaway Golf. Too ashamed to mention the drugs, he said he couldn't stop drinking, and he'd tried to kill himself. That's why he'd gone missing. The company arranged for a stay at an in-patient rehab facility, Scripps McDonald Center in San Diego, where Murray now works as a counselor.

He hasn't retained many reminders of his past; he says he doesn't do regret, and that his life now is an entirely new one. But he likes to burn CDs of his old band's music off his laptop and hand them out as a kind of free calling card. Murray was pretty good. He could wail in the best R&B style, and when he sings a cover of a blues number that B. B. King popularized, shouting "the thrill is gone, baby . . . I'm free from your spell," you have to remind yourself that the song is about a woman.

8

THE INFIDELITY PARADOX

We understand if you are, at this very moment, rubbing your chin and asking yourself, "If we're so addicted to each other that we're scared to death to be apart, how could anybody ever break up? And what explains adultery, and cheating on your boyfriend or girlfriend?"

Those are excellent questions, part of a very tough puzzle societies around the world have been struggling to solve for millennia. The question of why this or that relationship fails is unanswerable, partly because everybody's circumstances are so different. But there is a general truth about sexual relationships: As you'll see below, passion fades. Before it does, however, it can fill a multitude of lurking potholes. Once it's gone, the simple recognition that two people are ill-matched in personality or temperament probably accounts for many separations, and at least some of the roughly 43 percent of all first marriages in the United States that dissolve. Even then, though, breaking up really is hard to do, more testimony to the power of the addiction designed to keep us together.

Infidelity can be a wholly separate problem from personality and temperament clashes, though, of course, it, too, accounts for many dissolutions. While adultery may also have many ingredients, it is a universal phenomenon, even, as we've pointed out, in prairie voles. While prairie voles are socially monogamous, they're not nearly as strict about sexual monogamy as is popularly believed. Variations in brain circuits do exist and can significantly influence a vole's, or a person's, susceptibility to engage in sexual shenanigans.

This creates an inherent paradox in our conception of monogamy. Social monogamy and sexual monogamy can be two entirely different things. Yet most societies insist upon linking them so closely that we try to think of them as one and the same. For many, that's not the way it works. Fred Murray, and the difference between liking and wanting described by George Koob, illustrate this. Murray was married, loved his wife, yet took on a mistress. His mistress happened to be drugs, not another person, but as far as the basic brain circuitry involved, there's not much difference. He didn't set out to wreck his marriage, or his family. In fact, before it all came crashing down, Murray tried to keep his home life and his drug life separate from each other. Some of the people in his drug life had no idea where he lived, whether or not he was married or had children, or even what he did to earn a living. He bought old, run-down cars and used those to drive to drug-related meetings, often parking those cars away from his house so his wheels couldn't be tracked. Once, when a man knocked on his door in the middle of the night looking for drugs, Murray answered the door, saying, "What are you doing at my house? This is my house! You never come to my house," then shut the door. He viewed his family life—his territory, if you will—as a very different thing from his exciting, if destructive, relationship with drugs.

Most people who have sex outside their bonded relationship also wish to keep that experience wholly apart from their home life, from their social bond. Generally, breaking down the original relationship isn't the goal of sex outside it: over 60 percent of men who have an extramarital affair also say they never seriously imagined themselves doing so until it actually happened. Some of us are clearly motivated, or capable of being motivated, to have sex with partners outside the bond, yet most of us swear fealty to the bond, can be happy within it, and have no wish to part from it.

Hence the constant rotation of made-for-TV celebrity/politician/religious leader confessionals. TV preacher Jimmy Swaggart set the tone for these back in 1988 with a weepy mea culpa to his congregation and a vast television audience. He had "sinned against" Jesus, and pretty much everybody else who looked to his brand of moral guidance, after a rival evangelist photographed him with a Louisiana prostitute. To some, Swaggart got a justified comeuppance, since he'd loudly condemned yet another Christian leader, Jim Bakker, for his own sex scandal just the year before. Neither Swaggart nor Bakker had any desire to destroy their primary social bonds, yet both strayed from the path they considered righteous, driven by some force stronger than their moral conviction.

Just how many people sexually stray is open to debate. Despite the best efforts of social and scientific researchers over many years, none can say exactly what percentage have sex outside what is supposed to be a sexually monogamous relationship. As you might expect, interviewees are often reluctant to tell the truth in face-to-face interviews, and even anonymous written surveys are somewhat unreliable. Still, there are ballpark numbers.

In the years just before the much-vaunted "sexual revolution" of the 1960s, doctors in a New Orleans hospital conducted a survey of women admitted to the facility. The women were divided into two groups, those with cervical cancer and those without. More than half the women in the cancer group, 54 percent, said they'd had sex outside their marriages. That may not seem surprising, since cervical cancer is caused by a sexually transmitted virus, and so the more partners you have, the greater your chance of being infected. But one-quarter, 26 percent, of women *without* cancer also said they'd cheated on their husbands.

In the largest and most comprehensive poll of its kind, published in 1994 as *The Social Organization of Sexuality*, Edward Laumann and colleagues found that among American women born between 1943 and 1952—those who would have been in their forties and fifties at the time— nearly 20 percent reported they'd had sex with somebody other than their husbands. Just over 31 percent of men in the same age group had done so. Among nonmarried, but supposedly sexually monogamous, couples, like those who are living together or dating, cheating rates have been reported at upward of half.

THE INFIDELITY PARADOX • 213

Whether or not human beings are in a monogamous relationship, we do tend to covet thy neighbor's wife—or husband or boyfriend or girl-friend. A multinational survey of nearly 17,000 people, in fifty-three nations, from all parts of the globe, revealed that men and women, or at least the college-age men and women in the sample, were doing a lot of what social scientists and wildlife biologists call "mate poaching." Roughly half had made at least one attempt, and many of them succeeded. In North America, 62 percent of men, and 40 percent of women, had tried to poach somebody else's partner for a short-term fling.

A lot of people took the bait, too. Sixty percent of men who said they'd been a poaching target said they'd agreed to short-term sex with the poacher. Just under half of women said so. Similar successes were reported by poachers who said they'd made an attempt to steal a mate in order to establish a long-term relationship. Interestingly, societies in which women had more political power also had more mate poaching by both women and men.

Sex, of course, leads to babies. Right now, all over the world, millions of human babies are being unwittingly raised by cuckolded men. The exact percentage is unknown; results of studies vary wildly according to what region the study covered and other variables. A 1980 estimate from Hawaii came up with 2.3 percent. A study of the Swiss said 1 percent. One from Mexico said 12 percent. The best guess seems to be that between 3 and 10 percent of children around the world are being raised by fathers who don't know the children are not genetically related to them.

So, for the sake of argument, let's accept a rough estimate of 30 to 40 percent infidelity in marriage; 50 percent in nonmarital, but monogamous, relationships; and up to 10 percent of babies not being genetically related to the man who believes he's the father. All over the world. Among all races, and creeds, and cultures. The conclusion has to be that infidelity is an inherent behavioral trait for at least some part of the human population.

It has always been thus. One of the most famous speeches written by the ancient Greek writer and orator Lysias is a defense of a man who killed another man after finding his wife in bed with him: "I never suspected, but was simple-minded enough to suppose that my own was the chastest wife in the city," the accused man told the court.

Across time and culture, the paradox that wedges itself between social monogamy and sexual appetite has been a nasty problem. For thousands of years societies have tried to remove it, to force social monogamy and sexual behavior into alignment, usually by harnessing, corralling, and taming erotic desire. Marriage itself, the institutionalization of human love, is, in part, an attempt to provide structure and rules for sexual behavior. For many cultures influenced by Christian tradition, marriage serves to contain erotic reward within a social bond as a defense against original sin. Augustine of Hippo, who wrote and preached in the late 300s and early 400s, set the tone. Sex, he taught, was the result of man's fall from grace and banishment from paradise. He addressed the problem of illicit desire by arguing that, in Eden, sexual passion didn't exist in the same form as it does for postlapsarian people. It was wholly reined in by our rational selves. Orgasms weren't the source of vibrant, mind-blowing pleasure; they were sedate and of a piece with every other fine thing in the garden. Adam's and Eve's members were joined with no more fervor than "a slate to a slate." The burden of sexual desire and temptation was part of man's punishment for defying God. Shoving its unbridled wildness back into the harness of reason had to be one of man's primary duties to God if man was to ever regain paradise.

Given the debased form of human sexuality after the Fall, it would be much better to never have sex at all, some church fathers believed. But they also recognized that the weak willed would be viciously tempted to defy God's plan. So they gave them a way out. People could cave under the force of sexual desire, but only within marriage, and only under tightly regulated circumstances. Even in marriage, having sex out of lust or for pleasure was a mortal sin, and a wife was to guard herself against performing any act, any "whorish embraces," that would inflame her husband's erotic delight.

Violation of these rules was punished severely. You could lose your property, your family, your freedom for committing adultery. And yet, despite the repression, lots of people still committed adultery. No matter how terrible the consequences, they could not overcome the power of the human brain to tilt people toward behaviors they know can result in serious trouble.

So there was a "disconnect between what people advocated, and what they actually expected and tolerated," Stephanie Coontz, professor of history and family studies at The Evergreen State College in Olympia, Washington, tells us, echoing the disconnect between Murray and his home life and drug life, and the disconnect between a spouse who has extra-pair sex, but treasures the primary bond. "Moralists and philosophers usually lauded fidelity and condemned infidelity, but most of the time this was up in the same realm of abstraction as their endorsement of celibacy, peace on earth, and good will to all."

Coontz, who wrote the book *Marriage, A History: How Love Conquered Marriage*, explains that even during the strictest periods of sexual repression, accommodations were made. Many medieval European towns, for example, had a recognized, legal brothel. Among elites, there was an explicit recognition that marriage and romantic love—including erotic longing—were two different things. Romance was extolled as a higher form of love. "The cult of courtly love held that the truest love could only be found outside marriage," she points out. Indeed, the first rule of the twelfth century's *The Art of Courtly Love*, by Andreas Capallanus, was "Marriage is no excuse for not loving." "The only true love was adulterous," Coontz continues, "since marriage was economic and political and therefore not true love. You were marrying for practical reasons."

By Chaucer's time, in the late 1300s, European literature was full of comic, and not-so-comic, tales of adultery and cuckoldry. Malory's *Le Morte D'Arthur* revolves around the affair between Lancelot and Guinevere. Meanwhile, in the nonfiction world, church leaders struggled to contain licentiousness. Reading ecclesiastical accounts of the era, it can sometimes seem as if battling illicit sex had become the primary occupation of Christianity.

In the sixteenth and seventeenth centuries, men spoke and wrote freely, even to their fathers- and brothers-in-law "about their adventures among serving wenches, catching the pox from a prostitute, giving a good rogering," Coontz says with a laugh. They would talk about hiring a new serving girl who was good in bed, and have complete confidence nobody would tell their wives.

But the kernel of what the reluctant bridegroom Alfred Doolittle, in George Bernard Shaw's *Pygmalion*, disgustedly called "middle-class morality" had already begun by Chaucer's day. "Chaucer was an amazing social historian," Coontz says. *The Canterbury Tales*, a seminar in the various crosscurrents of Western society's struggle with inconvenient sexual desire, shows the beginnings of what most of us now consider ideal married life.

"The Franklin's Tale," told by a small landowner, a middle-class fellow, is about a middle-class couple: a sober, unglamorous working knight and his higher-status wife, Dorigen. In wooing his wife, the knight promises to not behave as a master toward her, to obey and serve her, if she will preserve the public image of his mastery in order to protect his reputation as a knight. She, in turn, swears to be his "humble, faithful wife." In other words, they enter a very modern compact of equality.

Another man, a squire named Aurelius, is, in contrast to the stable knight, a "servant of Venus." He falls for Dorigen. Aurelius pursues her constantly. He even threatens to kill himself if she can't give him some hope. Finally, in her husband's absence, and in an effort to save the lovesick squire while protecting her virtue, Dorigen promises to have sex with Aurelius, but only when the very landscape of the nearby shore changes, something she's sure will never happen. When the change appears to come to pass, a distraught Dorigen weeps out a confession to her husband, who lovingly tells her to honor her promise, but never to speak of it or associate it with his name. Aurelius, moved by the couple's dignity and love for each other, releases Dorigen from her vow.

"This was Chaucer's model of companionate marriage," Coontz explains. "This was a marriage that represented the emerging middle-class

THE INFIDELITY PARADOX · 217

values of partnership, and we start to get this glorification of companionship and mutual fidelity between husband and wife as opposed to the immorality of the upper and lower classes. That started the idealization of marital relationships and middle-class values. Chaucer was very prescient."

A key tenet of this vision of companionate marriage has been an otherworldly view of women and womanhood. In Chaucer's day, female erotic desire was taken for granted. In fact, three hundred years before Chaucer, and lasting at least until the 1700s, churchmen viewed women as cesspools of wicked temptation. Female desire was so fearsome, it gave rise to the myth of the *vagina dentata*, the vagina with teeth. But by "the nineteenth century," Coontz says, "the view that emerged of marriage was based on the idea that women were pure and virtuous."

A pure, virtuous woman who regarded sex as dutiful and who could not indulge in whorish embraces wasn't much fun in bed, though. So it was at least somewhat understandable that men might seek such pleasures elsewhere. Hence the famous "double standard." Men could see prostitutes, maybe have a mistress, but any such side action was not meant to be fatal to the marriage. Wives weren't supposed to protest. "I ran across a lot of diaries and letters, including ones where a woman did make an unseemly fuss about it, and her own relatives told her this was not a seemly thing to do," Coontz says of her research.

During the 1920s flapper era, and after books like our friend H. W. Long's, Americans and Europeans were reacquainted with the idea of female sexual longing. As women's economic power rose, and they acquired lives outside the home, they also encountered more opportunities to satisfy that longing, sometimes in the form of extramarital dalliances.

Facilitating cheating was certainly not the intention of the Dr. Longs of the world. Women's sexual liberation, Coontz points out, was supposed to be good for marriage. With the advent of the marital sex advice books, men weren't "supposed to have an excuse to venture outside marriage to get sex they couldn't get at home, because now they could."

This sexual liberation was aimed at the middle class. Lower classes weren't expected to be terribly "moral." The elite had license. The middle class was supposed to be the class of rectitude, the steel spine supporting the country.

Since the 1930s, Coontz believes, we've been living in an era that is more censorious of sexual infidelity than ever, even as sexual mores outside marriage have become more tolerant than ever. There is no longer any disassociation between eroticism and the primary bond. In fact, the primary bond, the marriage, isn't supposed to be disassociated from much of anything; we have exulted it to such a degree that we now expect it to provide all our happiness and fulfillment. Many marriages, Coontz thinks, crack under the strain of such high expectations.

We give you this quick scan of history to point out that for over 1,800 years of Western civilization, much of the world has been trying to come to grips with the central paradox of love—how it can coexist with infidelity. At first, sexual pleasure was suspect, often sinful, at the very least frowned upon, even in marriage. During the past hundred years or so, sexual bliss has become one of the main objects of marriage. And yet, regardless of how society has viewed marital sexual relations, both men and women still had sex outside marriage, and outside every other type of supposedly sexually monogamous relationship. This is because infidelity isn't caused by lax social mores or rigid social mores. A tendency toward extra-pair-bond sex can be programmed into our brains.

BLOCK THAT POACH!

Suppose you are a young account executive in an advertising agency. (We'll make you a man, but as Susan back in Minnesota demonstrated in chapter 2, given the estrogen rush of ovulation, the scenario works for women, too. Just switch the genders in this story.) One day, you step into the elevator for the ride up to the agency's offices and you see a beautiful woman. She's dressed professionally but alluringly, in high heels, a tight skirt, with loose, flowing hair. She's wearing glasses. You've had a thing about beautiful women in glasses since you were thirteen and looking at magazine pictures of sexy teachers. The attraction is immediate and powerful. You make eye contact and exchange smiles. The neurochemical events we've covered in earlier chapters start into motion. Oxytocin and vasopressin are released, dopamine trickles into the nucleus accumbens, and you are motivated to make a pass. You are, however, not a lab animal but a human being, and right now your rational brain is frantically pointing out that you've met her before, at your boss's house. She's his fiancée. Even if she were not, you're married, and while your sex life is rather dull, and the old urgency of your first years with your wife has dissipated, you love her. You hate the idea of losing her, and, brother, you'd definitely lose her if she found out you had an affair. You might also lose half your money and property, not to mention the price of a good divorce lawyer. Besides, you ate garlicky linguine with clam sauce for lunch. You settle for a friendly nod, a smile, and, when the elevator door opens, you retreat to your desk and sit down with an unconscious sigh.

This is self-control. Your prefrontal cortex has communicated with your amygdala, ventral tegmental area (VTA), and accumbens, and said "Cut it out!" You have just been confronted with a desire-reason dilemma and favored reason.

German neuroscientists Esther Diekhof and Oliver Gruber subjected test subjects to a desire-reason dilemma in order to explore functional differences in the PFC-accumbens link. First, eighteen young volunteers completed standardized written surveys of impulsivity and novelty seeking. Then, in phase 1 of the test, they played a kind of game in which they were shown one colored square at a time. They could choose to collect or reject each square by pressing a button. After each choice was made, they were informed whether that choice led to a small reward—one point—or not. Points equaled a money payout, so the more points they collected, the more money they won. This exercise got them acclimated to earning rewards by picking squares.

Subjects did the same thing in phase 2, but this time, they were being scanned in an fMRI, and, unlike the first round, they were forced to seek a long-term goal—collecting a predetermined block of three colors by the end of the game. If they succeeded, they'd receive a high number of points that determined the amount of money they would win. They were also free to choose a color that was *not* part of the target block. Choosing a color outside the block—"cheating" on the colors they were supposed to select—would yield a point and add to their overall final score. So they could earn more money by being daring. But it could also sidetrack them, and failure to meet the goal resulted in disqualification, wiping out any points. Therefore, it was safest to stick to the long-term objective and not yield to the temptation to rack up as many points as possible—to play it straight down the middle. In other words, they were given an incentive to seek an immediate reward, even as they were supposed to be focused on the delayed reward of collecting all the colors in the assigned block.

When the volunteers chose the immediate rewards, their brains displayed heightened activation of the accumbens and the VTA. When faced with the desire-reason dilemma, those areas were muted. There was negative signaling from the PFC, and it turned out that those people with the lowest impulsivity and novelty-seeking personality scores on the standard survey showed the most inverse signaling between the PFC and the accumbens and were most successful at staying focused on the long-term goal.

The scientists' results jibed with animal data, they said, providing the "first evidence that human beings may engage a similar mechanism, because they demonstrate an increased negative coupling between" the PFC and the accumbens and VTA "when desire collided with reason." Human beings are born with a bias toward satisfying immediate rewards, they argued, and our ability to counteract that bias in the service of a longer-term goal—like preserving a bonded relationship—when we sense something—sexual opportunity, for example—that sparks our reward circuits, may depend on the strength of the interaction between those circuits and the PFC.

As we've mentioned several times, there are a number of ways to weaken the PFC-accumbens link. Let's say it's now two years from the time you met the woman in the elevator. Let's also say she's no longer the boss's fiancée. And let's put you on a business trip to Columbus, Ohio, where you're staying at the Hyatt out by the airport, and instead of just smiling politely, the gorgeous woman with the glasses says, "Don't you hate these cookie-cutter hotels?" during the ride in the elevator. Then you say something about being bored because the only thing to do in Columbus is watch Buckeye football, and the next thing you know, you're in the bar of the cookie-cutter Hyatt sharing manhattans and laughing at each other's jokes, which aren't even all that funny, but the alcohol is muting your PFC so you're not being too particular. Her hand touches your shoulder. She looks directly into your eyes. A little oxytocin is released. A little dopamine trickles into your accumbens, and that sexual appetite that's gone missing in your marriage pops wide awake. You are not thinking of your wife or the price of a divorce lawyer.

Like people, prairie vole males and females will have sex outside the pair-bond. Yet we've spent a lot of words explaining that once males are bonded to a female, and their reward centers reorganized, males will attack intruder females. We've also said that most females won't bond to any other male if their partner dies or disappears. What, then, accounts for this paradox?

The male prairie vole's vasopressin circuitry modulates his attachment to territory and space and helps forge his monogamous bond. A strange female (or a male, for that matter) who wanders in is violating the resident male's homeland. He didn't care before being bonded, but now he will attack. If he is out wandering—on a business trip—looking for food to bring home to the wife and kids, and happens to run across a female who's been brought into estrus by another male, he may not be able to—or want to—resist. True, his brain has been changed with the bonding to his partner, and just as we described in the previous chapter, he's addicted to her. But that doesn't mean his appetite for responding to the sweet smell of a fertile female has been wiped out. Even if that fertile female is in his space, her aroma and his sexual urge can overwhelm his urge to protect his territory or think much about his partner. His appetitive reward system switches on. He's driven to mate. He isn't likely to invest time and energy in courting a female. He won't typically try to bring a non-estrous female into heat. But if she's already hot to trot, he'll be happy to have a fling, like the 60 percent of men who said they succumbed to a mate-poaching attempt by a woman. Afterward, the vole won't have a partner preference for her—that reorganization of his brain from liking to wanting still socially bonds him to his partner. Instead, his reaction is more like, "Oh, that was nice, thanks," and then he'll return home as if nothing ever happened.

The female side of the affair we've just described is equally casual. Her estrogen is up because she's been brought into estrus by the smell of her partner. But she's alone, now, out looking for food. Or she's sitting in the nest when her mate's away. A strange male comes along. Driven by her appetitive desire circuitry, she allows him to mount. Then she comes home, or waits for her male to come home, with no change in her relationship.

Each one of these "cheating" voles is still driven by the CRF stress response to maintain their bond and to return home. They still don't like being apart. They just happen to have had sex with other individuals. And because they have, a percentage of vole pups, just like a percentage of human babies, won't be the genetic offspring of the resident males in their nests.

Cheating voles and cheating people raise a question that's central to the paradox of monogamy. If it's sex you want, why not get it at home? Why take the risk?

Nature has played a trick on us. When a captive male marmoset is first introduced to a female, he becomes a very randy monkey. During the first ten days, he will have sex with his new girlfriend an average of over three times every half hour. The two form a monogamous bond. Sixty days into their relationship, they won't be having sex at all. But they will be huddling together much more than they did at first. Essentially, says University of Nebraska's Jeffrey French, who studied this pattern, within eighty days the marmosets "go from being young lovers to an old, married couple." They may not be having sex, but they have formed what, in Coontz's human terms, might be called a companionate marriage.

Reflective of lowered motivation for pursuing sex, the males' testosterone drops. So do their stress hormones. Their estrogen, meanwhile, rises. They've settled down.

"What happens when people get married?" Jim Pfaus asks. "Now that they can have sex any time they want, they stop having sex!" He's deliberately exaggerating, but it's true that the longer people are married, the less sex they have.

Sixteen percent of American married men age 40 to 49 said they have sex "a few times per year to monthly" in a 2010 national survey conducted by the Kinsey Institute for Research in Sex, Gender, and Reproduction. Only 20 percent of married men age 40 to 49 said they have sex "two-to-three times per week." Thirty-seven percent of married men age 25 to 29 (and presumably married for a shorter period of time) said the same. Women showed similar trends. Many factors affect sexual frequency and motivation, including children, work, bills, and how fit and healthy you are, but there's little doubt that the falloff is neurochemically based. Married men have significantly lower testosterone than single men, just like the marmosets. They have higher estrogen and lower stress hormones. They are bonded and settled. They give back rubs that are just back rubs. Living with somebody for a long time diminishes our interest in having sex with that person. It's sad, but true.

It's not that people, or monkeys, no longer like sex; they do. It's that sex with the same partner elicits a shrug, and their appetitive drive for sex overall has been diminished. They are less likely to seek sex with their own, or with a novel, partner.

There is a possible adaptive upside to a loss of sexual interest. Males that run around looking for sex aren't very good fathers. It may be that the chemical changes that accompany bonding help us pay attention to the job at hand: supporting babies. Also, when male guppies live with one female for a long time, they focus much less on sex and spend much more energy and attention trying to find food. As a result, they grow bigger and stronger than guppies who encounter a constantly changing cast of females. Those males put more effort into copulating and less into foraging for food, proving once again that being a playboy is a costly hobby.

This lack of sex drive that accompanies long-term partnership is one-half of a phenomenon discovered about fifty years ago and named, believe it or not, for the thirtieth president of the United States.

Calvin Coolidge is a comically unlikely avatar for any aspect of sexuality. Though he held office during the go-go 1920s, when the stock market boomed, hot jazz was the rage, and flappers bobbed their hair, he was a taciturn New Englander whose nickname, "Silent Cal," reflected his quiet manner and pithy wit. Today he's mostly remembered, when he's remembered at all, for two quotes. One is, "The chief business of the American people is business." It's possible the other is more apocryphal than factual, but it's stuck with him all the same.

As the story goes, Coolidge and his wife were touring a farm, guided separately by the resident farmer. As the farmer was showing Mrs. Coolidge around the barnyard, a rooster mounted a hen. The farmer was a little flustered by the display and, in an effort to ease his embarrassment by acknowledging the obvious, Mrs. Coolidge decided to ask a technical question: "How often does the rooster mate?"

"Dozens of times a day," the farmer replied.

Mrs. Coolidge smiled, and said, "Tell that to the president."

When the farmer took the president through the barnyard, and spotted the rooster, he dutifully informed him that Mrs. Coolidge had requested he pass along this fascinating bit of husbandry trivia.

"Same hen every time?" Coolidge asked.

"No, all different hens," the farmer answered.

"Tell that to Mrs. Coolidge," the president quipped.

About forty years later scientists were trying to explain a problem encountered by many labs that used rats as model animals. Male rats paired with a female had the frustrating habit of eagerly mating with her for a time, then slacking off until they became almost completely "unproductive." The scientists found that all they had to do to reignite the male's sex drive was put a new female in his cage. When they did, the depleted males became vigorous copulators once again. That's the Coolidge Effect, and in the fifty years since, it's been found to apply not only in rats but also in all mammals and in some animals as distantly related as hermaphrodite pond snails and beetles.

The first part of the Coolidge Effect, the slow death of passion, is experienced by many human couples. When physical passion disappears, there's less glue to bind them over the long haul of life, less excitement, less reward, often less closeness. If there are problems that had been buried under the passion, they can rise to the surface.

The second part of the Coolidge Effect, the rejuvenation of sexual appetite and performance, is a perfect example of the lure of novelty, and therefore an example of the lure of infidelity. It turns out that individual animals and people differ in just how powerful that lure can be when facing a conflict between desire and reason.

THE ATTRACTION OF "STRANGE"

Fred Murray loved novelty and sensation. His motivation to use drugs was so powerful it overwhelmed the bond he shared with his wife, and even his daughter. But the thrill of the new mistress eventually left him, too. The same stimulus, used in the same ways over and over, eventually blunts the dopamine effect, and addicts switch their motivation from liking to wanting.

With drugs, you can take more, and then more again, to keep getting high until your body finally gives out on pleasure's treadmill. You can't do that with a person if the thrill leaks out of a long-term committed relationship. Yes, you can try, but despite the shelves of self-help sex advice books, and decades of magazine articles passing along the wisdom of sex and love gurus, there are only so many sexy lingerie styles, only so many sexual positions, and only so many romantic getaways you can take to inject fire back into a relationship, and keep love's sweet urgency purring the way it did at the beginning.

But, without considering age, which also reduces our sex drive—also sad, but also true—living with somebody doesn't diminish our ability to enjoy sex itself. So we find ourselves well and truly bonded—addicted—

and less sexually interested in the person we are bonded to, yet still capable of being interested in sex with new people.

There appears to be strong genetic influence over our willingness to indulge this interest. In a large survey of 1,600 pairs of female twins between the ages of 19 and 83 conducted by Lynn Cherkas and coworkers in the United Kingdom, nearly one-quarter of individuals within the pairs said they'd had a sexual affair while married to, or living with, a partner in a socially monogamous relationship. Identical twins were 1.5 times more likely to have cheated than fraternal twins. Genetic inheritance accounted for about 41 percent of the difference, a strong link between genes and behavior.

Seventeen percent of the women who had cheated also believed it was always wrong to do so, yet, like Jimmy Swaggart, they did it anyway. Their moral belief, the scientists concluded based on their analysis, was engendered by the environment of their upbringing. So in 17 percent of the cases, the women defied the moral system they'd been given by family and education to follow some other, stronger, imperative.

Scientists like Pfaus, who study reward, have established that the Coolidge Effect occurs because the presence of a novel stimulus—a new sex partner—releases dopamine into the accumbens. The desire engine cranks over like a car getting a new battery, and the appetitive drive circuit sparks to life. The old spring is back. Rodents will start to seek sex. Humans will start working out at the gym, get new hairstyles, buy new clothes.

For the Coolidge Effect to take place, though, an individual has to appreciate the value of novelty and be daring enough to go after it. You have to be willing to leave the comfortable shelter of an established order— sometimes actually leave, as in leave home, and sometimes leave metaphorically, as in step outside the bond. There's risk involved in any such adventure. You may have to poach another individual's mate, perhaps over the objections of that individual, who may engage in some pretty serious mate guarding. The accused man helped by Lysias's speech murdered his wife's paramour. Sean Mulcahy, whose brother tried to poach his fiancée, loosed an angry outburst. In early 2012, a woman in Granbury, Texas, Shannon Griffin, was arrested for killing her husband's mistress in Kansas hours after she discovered the affair.

Animals mate guard, too, and so any would-be "adulterer" also faces risks. How willing they are to ignore those risks varies by individual. Take birds for instance. Zebra finches are small, colorful little birds native to Australia. They have orange-red beaks, white chests, gray back feathers, and black-and-white tails. The males have a splash of orangey feathers on the sides of their heads that make them look a little like the balloon-cheeked jazz trumpeter Dizzy Gillespie. Zebra finches are often used in scientific research, for a variety of reasons, but one reason they're used is that, like many bird species, they form lifelong monogamous bonds.

Yet some cheat on their partners. Using a captive colony of zebra finches, Wolfgang Forstmeier, along with a group of coworkers from the laboratory of Bart Kempenaers at Germany's Max Planck Institute of Ornithology, decided to find out what drives this extra-pair sex. Thinking in evolutionary terms, they figured the birds had to gain something by infidelity.

The male side of the equation might be easy to understand. Zebra finch males, like human males, have lots of sperm. By spreading that sperm around to many females, they are able to pass along genes to more offspring. Strict sexual monogamy limits that ability. The case for female wandering is sketchier. Unlike males, who don't have to use much personal capital to father chicks outside their own bond—males don't have to stick around to help raise them—females have to do the same amount of work no matter who fathers the babies. Yet they not only submit to strange males, they leave the nest to seduce them. One possible explanation has been that the females are looking out for their future chicks by seeking superior traits their long-term mates can't supply. So they try to pick winners by looking for very attractive males. There's some evidence that human and primate females do this, too, especially when they are ovulating. In fact, some primate females actually barter, exchanging sex for berries, or for meat, given by male sugar daddies—indicating a male is a good provider.

Forstmeier didn't exactly disprove the idea as much as modify it. He found that females may indeed favor bird Clooneys for extra-pair sex, but, at least in zebra finches, looks aren't everything.

After spending months watching over 1,500 captive finches, Forstmeier concluded that it was all about genes, all right, but not necessarily for looks. Some males were much more inclined to stray than others, and what largely determined that likelihood was the genetic inheritance passed to those Casanovas by their fathers, mainly for personality. Handsome males looked for extra-relationship sex, but so did homelier males. And the more a male, cute or plain, looked for sex, the more matings he had. So it wasn't just genes for good looks that were being spread; it was genes for novelty seeking, adventure, boldness.

After the team created a measuring system to gauge the female responses to these mate-poaching attempts, they found that female inclination to cheat was also linked to the genes *they'd* inherited from their fathers. Since promiscuous males—pretty or ugly—tended to get more matings, both sons and daughters of those males tended to have sex outside their bonded pair. The same traits that lent mating success to males also biased females more toward cheating, despite the fact that promiscuity might actually be costly to the females: the chicks of adulterous females tended to be slightly underweight. Nevertheless, Forstmeier concluded, it appears that promiscuity, even among bonded pairs, is to a significant degree a between-sex heritable trait. Certain individuals are just born more likely to seek sex outside a bond.

There's a lot of bird adultery. Female wrens take off before dawn to visit gigolos. About a third of the eggs laid by female savannah sparrows aren't fertilized by the fellow who's paying the rent.

A multinational team of European scientists announced in late 2011 that they had spent three years watching 164 wild great tit nests. Thirteen percent of the chicks resulted from extra-pair mating. Again, looks may have counted, but personality seemed to count more. "Bold" males sired significantly more chicks with females other than their primary mate than "shy" males, who tended to stick with their bonded partner. Bold female personality was also associated with having more extra-pair chicks.

Outgoing males were also more likely to be cuckolded themselves—while the boy's away, the girl will play. These personality types didn't seem to have much, if anything, to do with overall reproductive fitness: shy males and bold males fathered about the same number of chicks overall. Personality just influenced how likely each type was to have sex with somebody they weren't bonded to.

Researchers from Kempenaers's lab found an important gene allele that influenced how outgoing, gregarious, bold, and novelty seeking great tits could be. It was a variation of a key dopamine receptor gene, the bird version of the human D4 receptor, often referred to as *DRD4*, found mainly in the human prefrontal cortex.

In 2010, an interdisciplinary group from Binghamton University, in New York State, and the University of Georgia traced individual human personality differences to repeated sequences of the D4 receptor gene. People who carry one or more versions of the gene with seven or more repeats (7R+, sort of like the repeated sequences in vole *avprla* Larry found) tend to seek adventure, novelty, sensation.

The brains of those carrying the 7R+ repeats show differences in the way dopamine and its receptors are distributed, and the way these act in the reward circuitry and the PFC. For example, such people tend to have higher rates of attention-deficit/hyperactivity disorder, drug addiction, and alcoholism. They are prone to risking money, as in gambling or making chancy investments.

Naturally, nobody wants to become an alcoholic, or throw away their savings playing roulette. But the very same 7R+ repeats are also found in bold risk takers and novelty seekers who've proved vital to human advancement. People carrying this variant have tended to be the migrants, settlers of new lands. The allele also tends to make for restless innovators, the ambitious, the adventurous.

With all this in mind, the team surveyed 181 young adults to gauge their impulsive tendencies, intertemporal choice behavior, and sexual and relationship lives. Then they analyzed samples from each participant for the 7R+ repeats of the D4 receptor gene.

Those who carried at least one version of the 7R+ allele had 50 percent more instances of sexual infidelity than those who were noncarriers. The rate of sexual promiscuity was more than two times as great as it was in those who carried none. Half of the repeat allele carriers said they'd cheated on a partner they'd been monogamously committed to, but only 22 percent of noncarriers said they'd cheated. The 7R+ repeat carriers who cheated also said they had more extra-pair partners than the 22 percent of noncarriers who cheated.

It may be that monogamous populations require a certain amount of extra-pair sex to quickly fix beneficial gene variants into the gene pool. The group suggested that the D4 receptor gene is under evolutionary selection. When times are tough and scary, and the future uncertain, the population needs the bold and adventurous. In calm, plentiful times, there's less need. "That is, in environments where 'cad' behavior is adaptive, selective pressure for 7R+ would be positive; but in environments where 'dad' behavior is adaptive, selective pressure for 7R+ would be negative." Interestingly, the polygamous Yanomami Indians of the Amazon region are frequent 7R+ carriers.

The D4 receptor gene isn't the only dopamine receptor associated with the human ability to resist impulsive desire or yield to temptation for sensation and novelty. The reorganization we described in drug addicts and bonding prairie voles that switches the brain from liking to wanting, from an appetite for newness to maintenance of an existing relationship depends upon the interplay between two other dopamine receptors we've mentioned, D1 and D2.

Joshua Buckholtz, an assistant professor of psychology at Harvard University, has studied people with lower levels of D2 dopamine receptors in the striatum, the human brain region that encompasses the nucleus accumbens and is strongly linked to the amygdala, the PFC, and key dopamine-producing structures. He has found that lower levels of D2 in the striatum are associated with a propensity toward drug addiction.

"When faced with a novel stimulus or something salient for reward," he tells us, "such people don't have the machinery to tamp down this dopamine signaling. So more dopamine release causes an excessive drive to obtain the thing that elevated the dopamine signal. They're more juiced when they see a stimulus associated with a reward."

The stimulus could be money, food, drugs, or something erotic, but whatever it is, the PFC doesn't seem able to do its work, or is shouted down. These people, he says, tend to display high levels of impulsive behavior. They are also "more likely to violate the bonds of marriage, have promiscuous sex, and engage in risky behaviors more generally—to basically not adhere to the social norms relating to monogamy and cooperative partnership."

The story of dopamine receptors and what roles they play in which parts of the brain is extremely complex—Buckholtz describes the state of knowledge about how these play out in people as still "messy"—and so it's too early to start calling D4 a "cheating" gene or to say that people with lower levels of D2 receptors would make lousy spies because they'd be too susceptible to a Mata Hari. But while the close-up view may be fuzzy, the big picture is increasingly clear.

"The variation at the behavioral level" between one individual and another, Buckholtz says, "is due to variation at the neurobiological level."

CHEATING BY SOME COULD BE GOOD FOR ALL

Stephanie Coontz takes a skeptical view of what she calls "the hardwiring of human behavior." "I take a middle position on this. I think human beings are programmed for a capacity for both inclinations—toward monogamy, and toward promiscuity, or extramarital activities. Both desires are in us, both capabilities." She thinks social structures tip us one way or another.

Take the case of mate guarding. She points out that while mate guarding is ingrained in many human societies, and is a basic feature of monogamy, there's also a sociological explanation for it. "There does seem to be less mate guarding when there is less stuff at stake, or when survival depends on sharing rather than hoarding," she says, referring to Indians of the Amazon basin who traditionally practice a form of multiple social parentage with pregnant women having sex with several different men, all of whom are considered to have contributed something of their substance to the child, and therefore feel obligated to contribute to its support. "But once you have developed substantial differences in wealth and social status in a society, then there is a kind of narrowing of obligations."

Property and status are handed down to one's genetic offspring. Families weave connections based on genealogy. A bastard child is an intruder. "In my reading of history, families get really strict about female chastity when they do not want a child introduced to the family whose father or other relatives could make claims" on wealth or property, Coontz says.

Indeed, Roman Catholic priests were often married men until the First Lateran Council of 1123, when the church declared: "We absolutely forbid priests, deacons, subdeacons, and monks to have concubines or to contract marriage. We decree in accordance with the definitions of the sacred canons, that marriages already contracted by such persons must be dissolved, and that the persons be condemned to do penance." One of the reasons for this injunction—along with the old admonition against carnal pleasure—was Rome's fear that offspring would inherit church properties. Priests were "married" to mother church. It would brook no competition. Nuns had to remain celibate because they were married to Jesus. Any other relationship would be, in effect, adulterous.

Coontz's view isn't necessarily at odds with Larry's view that inborn neural mechanisms are at work. "Stuff"—wealth, property, kinship—is, after all, territory. And if, neurally speaking, men are babies to their female partners, it would be no wonder that women would want to keep an eye on them. Recall from chapter 2, for example, that men engage in more mate-guarding behavior when they sense, however unconsciously, that their lovers are ovulating. And mate guarding is a feature of all monogamous, and many nonmonogamous, animals.

As we've stressed many times, and as Coontz states, environment matters. For example, the *avpr1a* gene is very plastic: it responds to sociological conditions. When Nancy Solomon of Miami University in Oxford,

Ohio, placed voles into naturalistic enclosures for a period of time that roughly equaled how long a vole lives in the wild (typically around four years), she found that the *avpr1a* alleles the males carried "significantly influenced" the number of females they mated with and the total number of offspring they sired. This also meant that resident males were being cuckolded. Sure enough, bonded females gave birth to pups that were not the genetic offspring of their male mates.

Also using naturalistic enclosures, Alex Ophir traced a male's likelihood of extra-pair sex to vasopressin-receptor distribution in two brain areas, the posterior cingulate and laterodorsal thalamus, both of which are involved in spatial attention and memory. The males who were most likely to fertilize strange females had low levels of the receptor in these two areas.

And in the zebra finch study, about 28 percent of the chicks of the captive finches had discordant parentage. In wild samples, about 2 percent do.

Recall that in the large UK twin study, 17 percent of the women who cheated violated their own moral compass. Presumably, some number of women who might have been inclined to cheat but were more powerfully swayed by similar teachings did not. Had their environment not inculcated strictures on extra-pair sex, they might have relented to their natural inclination.

Culture is a reflection of our brains, often a reflection of the conflicts within them. Social bonding is certainly in conflict with sexual desire. And so we have had chastity belts, the burka, and female genital mutilation. We've instituted marriage and the enforced consequences of wrecking a marriage. Divorce isn't cheap; there's often a public shaming that accompanies a discovered affair, and there may be negative career implications. In the U.S. military, you could be subject to criminal charges for violating Uniformed Code of Military Justice provisions against adultery. These steps are society's way of trying to exploit our capacity for reason by adding costs to infidelity, and so constrain our desire for extra-pair sex.

The need for these bulwarks makes it appear as if humans are working at cross-purposes. Evolution may have built in such push-pull tension between social bonding and sexual drive. Over the millions of years of evolution, males and females may have been engaged in a kind of war of self-interest. Females may be constantly seeking the best possible genes for their offspring. To succeed, they would need to be both fertile and bold enough to exploit that fertility by seeking out an extra-pair partner. Meanwhile, males may be driven to spread all that sperm around but also be driven to prevent females from mating with any other males, especially during their fertile periods. So we jealously guard our mates and construct cultural norms of sexual monogamy to institutionalize the natural mate-guarding urge. We want monogamy for those we love but not necessarily for ourselves.

Throughout this book, we've seen how "variation at the neurobiological level" can make big differences in behavior: Frances Champagne's studies of rats, Todd Ahern's with the single-family voles, Larry's breakthroughs with vasopressin receptors and oxytocin, the sex-steroid influences that begin in our mother's wombs. Such variations can influence future relationships and sexual life, including the durability of relationships, and the propensity for having sex outside a pair-bond.

Kristina Durante, whom you met in chapter 2, has found that women with naturally higher levels of estrogen have reported that they're more willing to carry on a sexual relationship with a man who's not their primary, committed partner. They also appear to be more inclined to have serial monogamous relationships and to be on the lookout for a handsomer, richer, smarter guy.

Some people, such as those carriers of the 334 *AVPR1A* RS3 allele and those who have suffered through emotionally deprived babyhoods, may tend to form bonds that aren't as strong as those formed by other people and may tend to have more sex partners.

Other studies have shown that both men and women who have higher levels of testosterone are more likely to have sex with more people. This may be because a higher base level of testosterone blunts the first part of the Coolidge Effect, the drop in testosterone that correlates with a drop in mating effort, so they remain more motivated to seek sex.

Another way of putting it is that such variations moderate our individual capacities to fight with ourselves over our urges. Many find the idea that chemical bits of matter—our DNA, dopamine, the other neurosignaling molecules we've discussed—contribute so heavily to what we usually think of as morality as deeply unsettling, even offensive. But nature is neither moral nor immoral. It just is.

Evidence is mounting that nature has preserved a bias toward sexual infidelity in at least some human beings and that sexual adventuring is an inherent part of monogamous social systems. It may even be a *required* part of social monogamy. Yet humans often try to fully integrate our sexual drive into our social bond. But they aren't always easily married. Most of us treasure social monogamy, and most people—though not all— treasure sexual monogamy as part of that bonded relationship. Yet social monogamy dampens the erotic drive to have sex with that very social partner, while leaving us susceptible to seduction by another. Certain inborn differences in our brains can make us more welcoming to that novel stimulus—the pretty coworker, the handsome husband of a friend, the rich boss. Worse, the very sorts of people who most of us find attractive— the brave, bold, fun-loving, risk-taking sensation seekers—like the "cad" in chapter 2—the ones we usually call "a catch," may be the same people who are most biased toward having sex with people who aren't us.

In 2011, Dutch researchers announced the results of a survey on infidelity and intentions to commit adultery among a population of business executives. Of 1,250 respondents, just over 26 percent said they'd had an affair. There was a strong correlation between one's position in the corporate hierarchy and the likelihood of cheating, with those higher up the ladder, with more power, showing a much higher degree of both openness to an affair and actual infidelity. Gender didn't matter. The results held for both high-ranking men and women. These people were more confident, outgoing, and bold, which could account for how they managed to achieve career success in the first place.

This does not mean that every bold, impulsive risk-taking woman or man is going to be bedding new partners when they go on the road. Unlike animals, people have bigger, more powerful, rational brains. A lot of us are good at weighing risks and rewards.

But the very fact that most cultures have invested enormous resources in enforcing monogamy demonstrates how driven at least a subset of the human population is to have sex with people outside their bond. This happens across species. It's not just zebra finches, wrens, or prairie voles that cheat. Virtually all socially monogamous animals also have sex outside the pair-bond, including monogamous primates. Gibbon couples form strong socially monogamous bonds, but both males and females have intercourse with other partners. In 2005, when *March of the Penguins* (a documentary on emperor penguins) became an unexpected hit, social conservatives insisted that the incredible dedication male and female penguin partners showed in returning to each other after feeding—trekking across dozens of miles of frozen landscape—and in taking care of their young was a natural reflection of God's preference for monogamy and a lesson for all of us. Well, emperor penguins are indeed sexually monogamous, but only for one breeding cycle. Once the chicks can survive on their own, the family splits up, and the adults find new mates. It's as if Ozzie and Harriett Nelson switched partners with their neighbors Clara and Joe Randolph once Rick and David started school.

Some, such as Christopher Ryan and Cacilda Jethá, authors of a book called *Sex at Dawn*, argue the opposite extreme. They insist that sexual monogamy is wholly unnatural, purely a creation of human culture. This seems unsatisfying. Even if studies and surveys are off by ten, or even twenty, points, that would still mean that at least half of human beings who enter a long-term, sexually monogamous commitment don't have sex with any other person outside that bond. Likewise, not every prairie vole, zebra finch, or great tit engages in extra-pair mating, either. Only some do.

Also, many people express even more overall satisfaction with marriages that are decades old than they do with marriages younger than a presidential term, regardless of sexual frequency.

Social monogamy even seems to be good for our health, just like those guppies forced to live with one other partner. Married men live longer, and are healthier for longer, than single men. The same holds for married women.

The true answer to the question of whether humans are designed to be sexually monogamous seems to be, It depends. Some are. Others maybe not so much.

Sexual monogamy is less a matter of what people or animals are *supposed* to do than as it is of what they are, as individuals, *inclined* to do by their brains. Just because people like Swaggart, Bakker, and the famous-person-in-a-sex-scandal du jour might possess qualities that not only lead them to fame but also tend to bias them toward extra-relationship affairs does not mean that other people can't be more inclined to settle down into a cozy, happy life and never think seriously of having sex outside it, just as some people are more or less likely to be attracted to drugs.

Our millennia-long troubles are partially self-inflicted. Ecclesiastical scholars like Augustine taught that, yes, humans are *supposed* to be sexually monogamous. In fact, we'd all be a lot better off if we never had sex at all. They viewed the issue through the prism of natural law, arguing that God created human beings so that one man and one woman would forge an exclusive bond. Getting kicked out of Eden forced people to earn their way back by mimicking this original intent as best we could. We've been trapped in this one-size-fits-all dogma ever since.

Acknowledging a mistake does not mean that either social or sexual monogamy will die off. Rather, Coontz speculates, more couples will adopt new relationship modes. Some will feel best embracing both social and sexual monogamy, others will shuffle the deck. Some will negotiate sexual dalliances, others may issue a blanket don't-ask-don't-tell policy. "I also think that we should not discount the fact that as the age of marriage rises, people may have had twenty years of premarital sex, and when they marry, will say, 'I'm tired of this,'" and retire into happy sexual monogamy, she says.

"One thing I predict for sure," she continues, "is more diversity to satisfy all these conflicting desires." That's already happening as human beings try to reach a détente with the paradox living in our heads.

9

REWRITING THE STORY OF LOVE?

The hypotheses we've presented can seem pretty dark, can't they? Love is an addiction: not just a metaphorical addiction, but a real one. Some of us are simply inclined to have sex outside the social bond. Even those damn penguins down there on the ice cap, the ones in the documentary we all admired, aren't quite so monogamous as we thought. Worst of all, love is just chemicals stimulating neural activity on well-defined circuits, and not meant to elevate us to some kind of higher spiritual plane, but to lure us unthinkingly into reproduction, maximizing our evolutionary "fitness." It's all so base.

We've been hearing the objections since before we began writing. People mostly object to the idea of placing so much responsibility for human love on the tiny shoulders of molecules in our heads. They worry how that view could affect the way we see ourselves as human beings. When Kathy French discussed the role of hormones like vasopressin in a seminar for her undergraduate biology students, "lots of people were offended," she recalls with a laugh. "They said stuff like, 'To reduce this magical, emotional experience to mere hormones!' I mean, they really were offended!"

"In one sense, it is surprising how much resistance there is to the idea of a biological basis for behavior," suggests Paul Root Wolpe, a bioethicist who serves as director of the Center for Ethics at Emory University.

Still, we do understand the criticism. We could easily be accused of committing what the great cultural critic Neil Postman referred to as "scientism," what William James cautioned against as "medical materialism," and what others have referred to as "reductionism." But the basic idea that emotions, and the behaviors those emotions drive, are created in our brains is a very old one. "Men ought to know that from the brain, and from the brain only, arise our pleasures, joys, laughter, and jests, as well as our sorrows, pains, griefs and tears," Hippocrates wrote. Two thousand years later, T. H. Huxley less eloquently declared that "all states of consciousness in us, as in [animals], are immediately caused by molecular changes of the brain substance."

A mechanistic view does tend to give rise to gloomy thoughts, though. In fact, you could argue (and accurately, we think) that the science could break a trail to evil. We can already see the first sodbusters staking their claims.

"Women trust me more than ever!" reads a caption under a photo of a pretty untrustworthy-looking guy on a Web site for an outfit called Vero Labs. "Liquid Trust makes women want me," reads another, this time under a photo of an attractive woman in lingerie undoing a man's necktie. In an ungrammatical bit of ad copy that unintentionally reveals just how sketchy the product is, Vero Labs insists that "96% of women's choice of man is not based on how physically attractive the man is. It isn't based on how nice or rich the man is. It is based on a powerful internal feeling— TRUST."

As you can probably guess, that trust is instilled with a spray of oxytocin. Just spritz it on the way you would cologne, and the next thing you know, you'll have a better job, you'll make that big sale, and good-looking women in diaphanous nighties will volunteer to remove your tie. Unfortunately for would-be lotharios and used-car salesmen, Liquid Trust isn't good for much besides giggles. Even if it does contain any oxytocin (and we're dubious about that), spraying it on your skin or clothing won't do anything for you or anybody you meet. Beate Ditzen refers to this sort of thing as a "horror." Yet, as of this writing, you can purchase Liquid Trust on Amazon.com for just under $35 (people do buy it).

Commercial labs have seen the potential to make money off the science of social relationships. After some of Larry's work on vasopressin-receptor genes was published and the Swedish study on men and pair-bonding came out, a Canadian lab began to sell *AVPR1A* analysis for ninety-nine dollars. Women could now screen potential husbands for the "cheating gene." One scientist has promoted himself on television and in print with claims he can predict whether a man will cheat on his partners. He's developed a purported five-point test extrapolated largely from studies conducted by Larry, and such colleagues as Markus Heinrichs, as well as others in the social neuroscience field. Chemistry.com, a hugely popular dating Web site, promises to "send you free personalized matches with the potential to trigger chemistry." These kinds of businesses will only increase in numbers, and consumers will have to hone keen huckster radar—just as they must with regard to phony cancer cures, homeopathic "medicine," and energy-focusing crystals.

Such present-day "chemistry" fakery may be easy to detect, but the near-term future could bring more serious risks, as well as ambiguities. Prospective brides or grooms and fathers- or mothers-in-law could start insisting on a suite of premarital genetic tests related to neurochemicals like oxytocin, vasopressin, dopamine, CRF, along with their receptors. In addition to bundling the familiar "tall, professional, SWM" in personal ads, why wouldn't men proclaim "*AVPR1A* RS3 neg." as yet another selling point? Men and women routinely demand body types in their personals. Why not toss in genetic types? Why not casually drop one's oxytocin-receptor, estrogen, or testosterone status, one's dopamine power or mu-opioid quotient into the initial conversation along with what one does for a living? "Yeah," we imagine a woman saying, with a wave of her hand, and a hair toss, "apparently, I've got, like, a *jillion* oxytocin receptors in my ventral tegmental area."

Let's suppose something like the product being sold by Vero Labs actually works. Or, more plausibly, suppose somebody developed an aerosol spray to trigger release of your own oxytocin. Do you think bankers, stock brokers, and real estate agents would hesitate to use it? Under the influence of social-bonding chemicals, we might be more likely to believe that a dumpy house described as a "cute fixer-upper" by a perky realtor really does justify the half-million-dollar price. In the Zurich trust-game experiment we discussed in chapter 5, investors who received the oxytocin spray, but whose trust was subsequently violated by the trustees, continued to exhibit trusting behavior—even after being taken. As Heinrichs explains, with extra oxytocin, "you do not care about social risks."

Robert Heath may not have been able to cure patient B-19 of his homosexuality, but he foresaw the impact of the work he was doing. "What is more important than the function of the mind in terms of not only individuals but social groups?" he asked. "What is more important in terms of the future of mankind, of man's survival, than being able to regulate and control the mind?"

We're willing to give Heath the benefit of the doubt, to assume he intended these thoughts in good ways, not in the creepy ways we are today so often ready to imagine. But, in Heath's reality, who would get to decide how one's mind is controlled? How would we feel about a parent who medicates a child in an effort to encourage sexual abstinence or to nudge them toward being less anxious and more social? Millions of children in the United States, the majority of them boys, take Ritalin, a dopamine booster, every day, ostensibly to help them pay attention. But the drug also helps these children behave in more socially acceptable ways. Is this a model of the future?

Some people with Asperger's syndrome resent any implication there's something "wrong" with them that needs correcting. A few people with Asperger's believe themselves superior to those they regard as "neurotypicals." One person on the Asperger's end of the autism spectrum heckled Larry at a talk, arguing that Larry's research was an effort to make people like the heckler more like Larry. Any such treatment would only open those on the autism spectrum to all the social burdens other people face, the heckler suggested. "Who needs that?" he asked.

This argument echoes one frequently heard from ethicists: What if we cured genius? What would the world would be like if Beethoven, van Gogh, and Einstein were well-adjusted fellows with a family life out of

Father Knows Best? The price of genius often does come with antisociality, crummy relationships, and personal pain. On the other hand, while van Gogh's anguish might seem an acceptable trade-off to those of us who enjoy *Starry Night*, it was pretty lousy for van Gogh. So how should any possible therapies be offered? Who should get them? Who should control the use of brain-changing, emotion-altering drugs?

The nascent field of neuromarketing seeks to efficiently exploit the systems you've learned about here. In truth, all marketing is, on some level, neuromarketing—advertisers and product manufacturers have been trying to appeal to our emotions for hundreds of years. Pharmaceutical companies bolster their sales forces with attractive young women, and Cinnabon pumps the enticing aroma of baking buns into airport terminals and shopping malls so the sensory input from our eyes or our noses sets off a craving for reward, motivating a purchase. At the moment, neuromarketing is more slogan than science. But what if neuromarketers get really good at their jobs? At what point does a nudge become a shove?

What about the possible use of neurochemicals during interrogations of enemies? It might seem less a betrayal of American ideals than torturing people with waterboarding, but would it be ethical? In the past, the US Army Field Manual has suggested, in agreement with the Geneva conventions (which addresses the treatment of prisoners), that the use of *any* drugs not medically necessary is forbidden. But, according to a 2004 report by the Congressional Research Service, that restriction was slightly altered in a new edition of the Army Field Manual to forbid interrogators from using "drugs that may induce lasting and permanent mental alteration and damage." This appears to leave the door open for the use of neurochemicals.

THE GOOD NEWS

A big problem with any regulation or law that tries to cut off nefarious uses of science is that what can be scary can also be enormously helpful to many people. The man who heckled Larry about Asperger's was wrong to suggest that we think any competent adult should be required to accept a treatment. But he was correct that the work being done in labs like Larry's has the potential to create medical breakthroughs for conditions like autism spectrum disorders. "We are, by nature, a highly affiliative species craving social contact," Larry's former mentor, Thomas Insel, has stated. "When social experience becomes a source of anxiety rather than a source of comfort, we have lost something fundamental—whatever we call it."

It's true that autism has been "linked" or "associated" with many gene variants and environmental insults, so it's important to be cautious. But many hope that, by understanding the neurochemistry driving affiliation, social reward, and social bonding, we may one day be able to translate that knowledge into drug therapies to ameliorate some symptoms of autism. Toward those ends, Larry has established the Center for Translational Social Neuroscience at Emory University. As Maria Marshall's story shows us, conditions that prevail even at very young ages can affect us far into adult life. What if a young child were diagnosed with an autism spectrum disorder, then treated with something as simple as intranasal preparations of neurochemicals to push their brain away from a more severe spot on the spectrum? Such a therapy would probably never be a cure. But it might be possible to take advantage of brain chemistry to help make eye-to-eye contact and social interaction more rewarding for the autistic child. If such a child learned to associate eye gaze with reward, they might more closely attend to social cues. A snowball effect might take over, making the child better at reading emotions and less anxious about interacting with others. New, stronger neural connections could be established. The effects might last a lifetime. Larry believes drugs that activate the oxytocin system, combined with behavioral therapies, will one day do just that. Oxytocin is already being sold to parents of autistic children. In Australia, for example, parents are demanding, and receiving, prescriptions from doctors for intranasal sprays. This is unfortunate because, whereas experiments have shown behavioral progress in autistic patients given oxytocin, such improvements have so far been transient and modest.

More research is needed. In the meantime, parents may be buying into false hopes or, even worse, doing some harm from the uncontrolled way they're using the drug.

Melanocyte-stimulating hormone (MSH), which we briefly discussed in chapter 3, has some especially intriguing effects. This molecule (or a drug that could bind to its receptors, thus having similar action as MSH itself) can give you a tan. Companies in Australia are trying to market such drugs as agents that reduce risk of skin cancer. These drugs can also increase sexual arousal, decrease appetite, and stimulate oxytocin release in the brain, acting like vaginocervical stimulation in a pill. In other words, there could soon be a pill that gives you a tan, increases your sex drive, helps you lose weight, and activates the oxytocin system, presumably increasing trust, empathy, and bonding. Imagine the therapeutic boon to couples suffering a midlife crisis.

It also has potential as a drug to enhance social reciprocity in autism. Meera Modi, a researcher in Larry's lab, has shown that an MSH drug induces a pair-bond in prairie voles much more effectively than oxytocin does. This could mean that MSH-related drugs might also treat the social deficits of autism far more effectively than intranasal oxytocin.

Early, small-scale experiments on people suffering from social anxiety disorder, the most common mental illness after depression and alcoholism, indicate that oxytocin may be able to turn down the fear volume in the amygdala, making it easier for afflicted people to interact with others. In March 2012, scientists at the University of California, San Diego, announced that when a man experiencing social avoidance and relationship problems was treated with intranasal oxytocin, his libido, erections, and orgasms all improved, and so did his relationship. In tests, Parkinson's patients, who can lose some of their ability to discriminate emotions (either through the action of the disease or as a consequence of treatment), have benefited from oxytocin.

Schizophrenia can be especially intractable. Patients with that condition tend to have abnormal levels of oxytocin in their blood. When

246 · THE CHEMISTRY BETWEEN US

schizophrenics were given intranasal doses along with antipsychotic drugs, their symptoms improved slightly when compared with those who took only the antipsychotics.

Some researchers are now exploring the possibility of altering the brain's reward circuitry as a way to help addicts stay off drugs. Perhaps CRF action could be tamped down so recovering addicts are less likely to experience the negative motivation that urges them to return to drug use.

Associations have also been drawn between different receptor-gene variants and behavior patterns that don't really fall into the category of disorders but might make important differences in the lives of those who have them. One oxytocin-receptor variant has been associated with less empathy in mothers toward their children. Another, which has been linked to emotional deficits, appears to be influenced by environment. Girls who have this variant, along with some negative childhood experiences (a mother with depression, for instance), are, in turn, more likely to suffer depression and anxiety. A vasopressin *AVPR1A* variant has been associated with the age at which girls first have sex. As we've seen, early-life stress can bias girls toward having intercourse sooner. Boys who have two copies of the long-RS3 version of *AVPR1A* tend to have sex before age fifteen, when compared with boys who have two copies of the short version.

A mother suffering from postpartum depression and anxiety might benefit, and her baby might as well, from treatment with neurochemicals. Distant fathers could show improved nurturing behaviors, too. Some psychologists and scientists are now discussing the use of intranasal oxytocin as part of couples therapy, in the wake of Ditzen's study of couples communication under its influence. Therapy itself is stressful. Since oxytocin dampens the anxiety response and tips the brain toward trusting, it could become a very useful tool for promoting open, positive communication between troubled partners. A psychiatrist might use neurochemicals to ease communication with, and compliance by, a client. If a patient took a dose at the start of a session, they might be more forthcoming in revealing their thoughts and motivations. This might benefit both the patient and the psychiatrist (and save billable time typically devoted to the small talk used to gain patient trust). Of course, any such use should be very closely controlled. When men were asked to complete a survey of their innermost sexual fantasies, then put the questionnaire in an envelope and hand it to a researcher, 60 percent of the men who received an intranasal dose of oxytocin didn't seal the envelope, thus leaving its contents accessible to the experimenter. Only 3 percent of men who received a placebo left the envelope unsealed.

THE CHALLENGE FOR SOCIETY

As William James wrote, "Something definite happens when to a certain brain-state a certain 'sciousness' corresponds. A genuine glimpse into what it is would be *the* scientific achievement, before which all past achievements would pale." Every new scientific or technological endeavor forces society to respond. The field of social neuroscience, and especially the study of bonding between people, should prompt serious reexaminations of habits, institutions, and systems large and small.

The character of nations can depend on how well they address the issues raised. In 1949, at the height of the Cold War, Geoffrey Gorer and John Rickman published a book called *The People of Great Russia: A Psychological Study*. In it, they argued that the Russian habit of tightly swaddling infants is "extremely painful and frustrating and is responded to with intense and destructive rage which cannot be expressed physically." In combination with other elements of Russian culture, like public shaming, this led to aggression and support for strongman leaders. Gorer and Rickman, their book, and the swaddling theory were all criticized—even laughed at. Such a response may or may not have been justified (though the theory was still being discussed in textbooks through the 1970s), but the concept that early life and genetic variation conspire to create a trend toward one style of behavior or another is now obvious, even on a national level.

Through research conducted in South Korea and the United States, scientists found that the Koreans lived up to stereotype: they were more likely than Americans to suppress their emotions. Then subjects in both countries were typed for oxytocin-receptor differences. Koreans with one type suppressed emotions more than their countrymen with a second type. Americans with the first type suppressed emotions *less* than Americans with the second type. The mirror-image difference was accounted for by the ways in which the cultures affected gene expression.

New knowledge presents us with the opportunity to think about making changes on both individual and culturewide scales that can affect our society. Some changes might be easy and small, but still very important. What if we changed the way we deliver babies? Lane Strathearn, of Baylor University, worries about the prevalence of cesarean sections. As we stated when we discussed how ewes bond with lambs, a cesarean bypasses the vaginocervical canal. This likely leads to significantly less oxytocin release into a mother's brain, possibly affecting the mother-infant bond. Few doctors or expectant mothers seem to be taking this into account when they schedule surgeries that aren't medically necessary.

The hospital-delivery experience itself could interfere with mother-baby bonding, Strathearn fears. "What do we do as soon as the baby is born?" he asks rhetorically. "We whip those babies away from their mothers rather than allowing longer skin-to-skin contact, enabling breast-feeding."

In a study published in late 2011, researchers found that newborns separated from their mothers experienced a 176 percent hike in autonomic activity (an indicator of stress) and an 86 percent drop in quiet sleep than when they were in contact with their mothers' skin. We're not suggesting that mothers forgo the hospital; there's no disputing the advantages of modern technology in preventing newborn and maternal injuries and deaths. But the habit of removing babies from their mothers so soon after delivery may be affecting both mothers' and babies' brains in detrimental ways, increasing, for example, the respective risks of postpartum depression and negative childhood behavior.

Sue Carter has raised another troubling issue. Anti-oxytocin drugs are sometimes given to expectant mothers experiencing premature labor. And, of course, oxytocin itself is given to mothers to induce labor. Vole experiments have indicated that interfering with these systems could create brain changes that influence later behaviors, possibly tipping the brain toward disorders like depression, anxiety, and even autism. While there isn't any human clinical evidence suggesting that oxytocin drugs used to manage delivery increase the risk of later-life psychiatric issues, the possibility does warrant investigation.

Any mention of parental behavior and autism risk is fraught with heated controversy. When Strathearn discusses the subject, he punctuates his half of the conversation with heavy sighs. After one such sigh, followed by a long pause, he says, "Among colleagues here at the hospital, I must be extremely careful how I word these . . . These are opinions." He's thinking of Leo Kanner. The founder of the psychiatric clinic at Johns Hopkins, Kanner used words like "frigid" to describe mothering styles. That sort of vocabulary was turned into "refrigerator mother," and an unfortunate tradition of mother blaming ensued. "Today, in the autism field, any suggestion that mothering may have any impact on the development of autism is rejected in aggressive ways," Strathearn explains. "So it is a topic to be very careful about, but I feel strongly we cannot ignore it, because autism is something we cannot ignore."

He believes that both mother-baby bonding and maternal nurturing styles do indeed play a role in the trajectory of autistic behaviors in children whose genetics and/or in utero development predispose them to autism spectrum disorders. "There is undeniable evidence out there, in both animals and humans, that social environment influences the development of social behavior in children," he argues.

It's possible that the social connection between parent and baby is something like physical exercise. Each time the infant and parents lock gazes, physically touch, or mutually smile and coo, the baby may be increasing its resilience against genetic or environmental risk factors for autism by strengthening the neural connections of the social brain, steering it toward healthy development.

As Strathearn fears, the nearly reflexive response of some to this kind of talk is anger. He's not blaming mothers or fathers, but pointing out that human interactions affect the brain, that the brain affects human interaction in a feedback loop, and that the behaviors of parents are an ingredient in what appears to be a complex stew affecting the trajectory of autism, or the development of the social brain.

As the kind of work Frances Champagne and others are doing has shown, stress and anxiety, especially in early life, can alter behavior far into adult life. That behavior can be transmitted to the next generation. Studies have shown that levels of neurochemicals like oxytocin and vasopressin are shared between human parents and their children and that the behaviors of both generations are associated with those levels. Parent-child pairs with lower levels of oxytocin engage less often and less rewardingly than pairs with higher ones.

Information like this ought to make us pause and attempt a panoramic view of our culture. We've been busy building a pretty anxious one. In doing so, we may be changing the collective social brain.

For example, the economy may not appear to have much to do with love, desire, and bonding. But consider something Strathearn often does. We know he thinks motherhood might be getting off to a bad start in the United States and other developed nations—but not just because of what goes on in our hospitals. "The mother brings the baby home from the hospital, and soon goes back to work and puts the child in day care." If you look at our world through the lens of bonding, Strathearn says, that may not be such a good idea. "You look at our society and the patterns we've created. We think we are making our lives better, but are we? Or are we creating perhaps more subtle—or not so subtle—problems?"

Mother-infant bonding is the keystone of all human bonding. But in our current economic world, many parents, single or not, have little choice but to return to work as soon as possible after giving birth. Stay-at-home parenting has become a luxury, and it's not just because parents want a ski boat or two weeks at Claridge's in London. It's because we're being crushed by health-insurance premiums, by caring for our parents, by college tuition costs, by fears of unemployment, by a changing workscape in which if you snooze, you lose.

Arguments over the economy and family life have been going on since the 1970s, but research into such questions has traditionally fallen into the realm of social science, making it vulnerable to accusations that its findings are "squishy." Now, though, social neuroscience is producing hard data that explain actual mechanisms behind how the emotional connections between parent and infant affect the developing brain, then go on to influence the next generation. In rats, we know how this happens—right down to the molecule.

Few people recognize just how important this research could be. Policy makers, politicians, and lobbyists are stuck in a past time, tossing out "personal responsibility" and then advocating for severe budget cuts, even for effective programs that intervene in the cycle of negative parenting. Such a cut might save a penny now, only to rack up a wave of costs later. It's fine to argue that a teen mother only has herself to blame for bearing a child she's not equipped to raise, that she should buck up and show responsibility. But such tropes assume perfect, rational control. As we've seen, there's no such thing. Like it or not, some will fail, and then society will bear the expense of future difficulties encountered, or caused, by a child raised in an emotionally or physically deprived home.

It's possible that the very culture of communication we've spent the past fifty years creating and congratulating ourselves about could be contributing to an alienated society because it bypasses the neural circuits meant to foster communal love. Too little direct human-to-human stimulation of these circuits can stunt their development. E-mail, texting, Twitter, Facebook—digital worship in general—have led to less face-to-face contact, all the while promoting the impression that technology can mimic the physical presence of human beings in time and space. We buy groceries in self-serve aisles, bank online or via ATMs, shop on our computers. We've created what Postman called a "Technopoly."

These trends could affect the way our brains work. The same Wisconsin lab that has studied adoptees from foreign orphanages ran a stress test on young girls. The researchers assessed the relationships between the girls and their mothers, and then had the girls perform a math and

252 · THE CHEMISTRY BETWEEN US

language test designed to increase their anxiety levels. The girls were divided into four groups. One interacted with their mothers in person. Another spoke to their mothers on the phone. A third texted with their mothers, and a fourth had no communication at all with theirs. Throughout, the scientists monitored levels of oxytocin in the girls' urine and cortisol in their saliva. Even after controlling for relationship differences, the girls who interacted with their mothers face-to-face had increased levels of oxytocin and less cortisol release than the other groups. The girls who used text messages to communicate with their mothers had no additional oxytocin release and higher levels of cortisol—the same as the girls who had no communication with their mothers at all.

A hypothesis advocated by Dutch scientist Carsten de Dreu, among others, sees oxytocin's functional evolution in light of groups. The first group consists of a mother and her infant. Another group is the human couple, then the immediate family, the extended family, the clan, the tribe, and so on. This accounts for the stunning success of human beings in not only managing to avoid extinction, but in dominating the earth. In-group trust, de Dreu proposes, is moderated by oxytocin and its relevant neural circuits. It provides the social lubricant—not just for the kinds of one-on-one interactions we've focused on here but also, when writ large, for entire societies.

Indeed, it appears that when people cooperate with each other, both oxytocin and vasopressin help build community trust. This was recently shown on a small scale by Larry's Emory colleague, anthropologist James Rilling.

In the years after World War II, as the threat of nuclear war rose, two researchers at the Rand Corporation, Merrill Flood and Melvin Dresher, studied game theory in an effort to figure out how two nations might react to various possible nuclear scenarios. They created what later became known as the "Prisoner's Dilemma." Imagine two crooks, in jail for bank robbery. They're held in separate cells. Police tell each one that if he cooperates, and the other prisoner doesn't, the snitch will get probation, and the accomplice will get five years in the penitentiary. If he doesn't cooperate, but his accomplice does, *he'll* get the five years and the other prisoner will get probation. If both men cooperate and confess, each will get two years in the penitentiary. If neither cooperates, they'll both get off with probation for a minor offense because the cops won't be able to prove a more serious case. If you're one of the crooks, what do you do? Well, it depends on how much you trust your fellow crook to clam up.

A version of the game can be played with money payouts, and that's how Rilling did it. The amounts of the payouts depended on levels of partner cooperation or partner defection. Intranasal oxytocin tended to increase cooperation. But that's not all it did. When men cooperated with each other, Rilling found, the oxytocin accentuated the activity of the striatum, reminiscent of its effects in the nucleus accumbens of voles during bonding. This effect likely increases the reward of reciprocated cooperation, facilitates learning the lesson that another person can be trusted, and that it feels good to trust. Both oxytocin and vasopressin increased levels of collaboration (though vasopressin increased it only if one player first made a gesture of trust toward the other), suggesting that the molecules may help build community trust by acting in specific brain regions, including the amygdala. This prompts a question: What happens in societies when there's little personal interaction among people other than within groups of close friends?

Violence and the environment are two other macro concerns. When Larry spoke at the annual Blouin Creative Leadership Summit, held in conjunction with the UN General Assembly meeting in September 2011, he argued that government leaders making policies for war-ravaged regions like Iraq and Afghanistan need to consider how stressful experiences in early life affect the brain and later behavior. It's not news that violence and neglect are bad for children. Harry Harlow's famous experiments in the late 1950s showed just how disturbed babies could become if they weren't nurtured or held. And many experiments, surveys, and life histories spanning many years and many places provide heaps of evidence that war, gang violence, and trauma come with severe consequences for the young people who live through them. Now that social neuroscientists are zeroing in on the brain mechanisms to help explain later adult behavior in light of these experiences, however, war planners must consider what kinds of blowback they may face when this young, troubled generation grows into adulthood.

Knowledge of the chemistry involved in the sexual differentiation of our brains should also make us consider how we manage the natural environment. Pollutants known as endocrine-disrupting chemicals (EDCs)—found in certain plastics, weed killers, even drugs—may be doing more to change our social brains than any other factor, playing the same role in the sexual organization of our brains as estrogen and testosterone used in the experiments conducted by Charles Phoenix and his successors. Among the most famous and widespread are BPA (found in can epoxy linings and heat-sensitive store receipts); phthalates (found, well, everywhere, but especially in softened, pliable plastics); atrazine (the nation's most popular herbicide, used on the majority of the US corn crop); and estrogens from drugs such as birth control pills. But there are dozens more. Experiment after experiment has proven that, even at low doses, exposure to EDCs in utero or in very early life can permanently alter the gender-typical behavior of lab animals, usually feminizing males.

At this stage, no one, including us, can say for sure what our new knowledge means for the future. But we think much more consideration ought to be given to the kind of culture we might be creating by actions, laws, and policies that may seem to have nothing to do with our social brains but that could have wide-ranging and profound effects on them.

WHAT IS LOVE? WHO ARE WE?

When Copernicus overturned about 2,000 years of thought by declaring Earth as one of a number of planets circling the sun, and as the Copernican view expanded with new discoveries placing our solar system in the Milky Way and the Milky Way in one of many millions of galaxies in an expanding universe, humans were forced to rethink the idea of their home planet as the center of things. Then Darwin forced us off our self-referential pedestal. With every such step, religious, social, and personal dogmas are forced out of the easy chair of belief into another, much less cozy position. Social neuroscience is now posing that kind of uncomfortable challenge to the ideas we started with: the ways we think about love, and how the ways we think about love make us think about ourselves.

We've tried to answer lots of questions in this book. But there's one question we can't put to rest: Why do we love? Science can never answer the big whys of life. Whys lead us down rabbit holes as any parent of a three-year-old peppering them with whys knows well. When junior asks, "Why do we fall in love?" we answer, "To have children." "Why do we have children?" And right about there, we find ourselves invoking something like either "God's plan" or "To share our love." This then opens us up to "Why do we want to share our love?" and then we unholster our lame bailout option: an offer to watch *Sponge Bob*.

Addressing whys are what religion, philosophy, and myth making are for. So we tell stories to help us make sense of the world and the universe. We use our stories to justify our own particular views.

William James understood the power of stories. In fact, it's often been said that Henry James was a novelist who wrote like a psychologist, while his brother William was a psychologist who wrote like a novelist. Though a man of science, William regretted the way new insights were sometimes used to undermine myth:

> Perhaps the commonest expression of this assumption that spiritual value is undone if lowly origin be asserted is seen in those comments which unsentimental people so often pass on [to] their more sentimental acquaintances. Alfred believes in immortality so strongly because his temperament is so emotional. . . .
>
> We all use it to some degree in criticizing persons whose states of mind we regard as overstrained. But when other people criticize our own more exalted soul-flights by calling them "nothing but" expressions of our dispositions, we feel outraged and hurt, for we know that, whatever be our organism's peculiarities, our mental

states have their substantive value as revelations of the living truth; and we wish that all this medical materialism could be made to hold its tongue. . . . Medical materialism finishes up St. Paul by calling his vision on the road to Damascus a discharging lesion of the occipital cortex, he being an epileptic.

Conservative philosophers, political theorists, and bioethicists worry about this very thing, and they are right to do so. They recognize, as many others don't, that the real pillars holding up our culture are not science, technology, factories, or even laws, but the stories we tell ourselves. They can be fragile, though. We sometimes make mistakes when we try to erode them—witness John Money and his insistence that society makes gender. So, in an effort to give them weight and authority, conservatives often try to build a kind of ethical Maginot Line by codifying narrative into "natural law," a set of immutable truths. Like the actual Maginot Line, though, natural law is itself flimsy. Far from being immutable, it changes with time.

Love is a good example of how this happens. If the story of any human thing would seem to be immutable, you'd think that thing would be love. As soon as we invented writing, we began composing pieces about love and desire. A Sumerian cuneiform poem from about 4,100 years ago begins, "Bridegroom, dear to my heart, Goodly is your beauty, honeysweet." The narrator goes on to ask this bridegroom to hurry up and take her to the bedchamber. He obliges, after which she says, "Bridegroom, you have taken your pleasure of me. Tell my mother, she will give you delicacies; my father, he will give you gifts." Some traditions may have changed (though we think this idea of giving the men some prosciutto and a Rolex after sex deserves a comeback), but the basic sentiment is recognizable to any of us living in any society these many generations later. There's the anxious anticipation, the sheer joy and eroticism, the contented afterglow. And then comes the story. Or the painting or the poem or the movie.

Yet we've changed our own view of natural law as it applies to love. In the United States, antimiscegenation laws, like slavery and the denial of women's political rights, were based on some people's interpretation of natural law to the effect that race mixing was abhorrent and unbiblical and would taint the white race. This view held in some states until June 12, 1967, when *Loving v. Virginia* was decided in favor of the appropriately named couple. Likewise, most Americans now think that not only is homosexuality not wrong, but gay men and women should be able to legally marry. This is a startling turnaround from our national belief just a few years ago, when the view that such unions were unnatural prevailed.

Science can take some of the credit for this. It has provided new information that can be incorporated into new explanations for the lives of people who don't fit into a strictly binary sex and gender world. But this changing tale has undermined the firmly held narratives of others. Some react to the resulting disorientation with denial. Others respond by lashing out in fear.

In the fall of 2011, Dr. Keith Ablow, who bills himself as "one of America's leading psychiatrists," and who serves as the Fox News expert on psychiatry and contributing editor to *Good Housekeeping* magazine, advised all parents to keep their children from watching *Dancing with the Stars* during the season that included Chaz Bono as a contestant. Why? Because Chaz Bono underwent gender-reassignment surgery to transition from being bodily female to being bodily male. Ablow wrote for Fox that "many of the children who might be watching will be establishing a sense of self which includes, of course, a sexual/gender identity."

If you think this sounds a lot like Money's old theory that gender can be imposed by society, you're right. "Human beings do model one another—in terms of emotion, thought and behavior," Ablow continued. "Chaz Bono should not be applauded any more than someone who, tragically, believes that his species, rather than gender, is what is amiss and asks a plastic surgeon to build him a tail of flesh harvested from his abdomen."

In other words, according to Ablow, watching Chaz Bono do the rumba could make your sons want to cut off their penises. These beliefs are not only fear based but also display astonishing ignorance. Yet, Ablow's willful denial of reality has traction.

When President Obama selected Amanda Simpson, a transsexual and former test pilot, as an adviser to the Department of Commerce, social conservatives were outraged. The American Family Association, an evangelical political-action group, called the appointment a "travesty." "What the deviance cabal wants more than anything is society's approval for their sexually aberrant lifestyles," it declared. Equally strident denunciations were issued by other influential right-wing political groups, such as the Family Research Council. Some of them mistakenly conflated the transgendered with homosexuals.

In 2011, when transgendered people petitioned New York City to let them alter their birth certificates to reflect their identified gender, even if they hadn't undergone transitional surgery, Peter Sprigg, a policy researcher at the Family Research Council, told the *New York Times* that any such change would be a form of "fraud."

"I think you have the objective reality of their genital makeup and their chromosomal makeup, weighed against the entirely subjective experience of so-called gender identity," Sprigg said. This placed Sprigg and his group in the same camp as Germaine Greer, an unlikely alliance indeed.

Transgendered people don't just wake up one day and decide it might be more fun to be a man or a woman. You can pretend you aren't homosexual, which can lead to trouble if you're a prominent male and have built a reputation for being antigay, only to find yourself hiring a male prostitute or picking up a man in an airport bathroom or trying to seduce young men in your congregation. You can have some homosexual experiences and not be a homosexual. But you can't be recruited to be a homosexual, and you certainly can't be "cured" of it if you are one. Heterosexual boys and girls behave the way they do because their brains are telling them to. No matter how compelling the TV commercial, toy makers can't force children to buy a gendered toy. Their TV commercials are tailored to make children believe a particular toy will fulfill the gender-based desire generated in their brains.

There is powerful interplay among culture, genes, upbringing, and our brains. But culture doesn't create gender—it reflects it. Gender influences everything, from who we love to whether we'll be more or less likely to think it would be really fun to turn the headboards of our beds into the top turnbuckle of an imaginary wrestling ring and take a dive off of it. (There's a reason why boys account for a large majority of injury-related emergency-room visits to US hospitals.) Yet, many people still argue that culture creates sexuality because that story fits their worldview.

Likewise, as we wrote at the start of this chapter, a mutating view of love also threatens many (perhaps most) people's story about the most important emotion of their lives. So how would understanding the neural systems of love affect our view of it? Could love that can be created on demand still be "real"?

There's a school of thought within neuroscience that suggests free will is a myth, that the preconscious brain informs the conscious mind, which then acts *as if* it's making a decision—when, in fact, our course of action was already decided before we were even aware of it. If true, some worry that love could be like quantum physics, in which the mere act of observing two entangled particles changes their characters. If we became aware of the brain mechanisms of love, could we really wreck it?

Our old friend Eduard von Hartmann, that Teutonic gloom-meister, thought so. Knowledge strips away the required madness and delusion, he believed. We have to be blind, he wrote, because when one "perceives the absurdity of the vastness of this impulse . . . it now enters into the passion with the certainty for *its* part of doing a stupid thing."

A good view might also make us think of love the way we think of car engines, slot machines, or computer software: as a programmatic process. In turn, we might come to see ourselves as programmatic as well.

We disagree with Hartmann. If he were correct, you could cure a drug addict by explaining the brain mechanisms of addiction. Sadly, it's not so easy. A meth user who understands the detailed molecular mechanisms of getting high, and the role of stress hormones, is still going to be a meth addict. Likewise, a person who's falling in love may understand the neural processes, but he or she is still going to be a sucker for love.

Whether free will exists at all—or it contributes 10 or 20 or 30 percent to our behaviors—is less important than the fact that we act *as if* we have it. In other words, we tell ourselves a story. *That* is what it means to be a human, especially one in love. It's why both of us believe love's future can be as bright as ever.

How, for example, would we react to the world depicted by satirist George Saunders in his short story "Escape from Spiderhead"? The narrator, Jeff, and a young woman named Heather meet each other in a lab where they're serving as guinea pigs for a new drug. After another character, Abnesti, who is administering the tests, gives them the drug, Jeff thinks Heather looks "super-good," and she thinks the same about him. "Soon, there on the couch, off we went. It was super-hot between us." And not only hot, but "right." Jeff and Heather think they're in love. Later, Jeff is given another drug, and the love disappears.

"That is powerful," Abnesti tells Jeff. "That is killer. We have unlocked a mysterious eternal secret. What a fantastic game-changer! Say someone can't love? Now he or she can. We can make him. Say someone loves too much? Or loves someone deemed unsuitable by his or her caregiver? We can tone that shit right down. Say someone is blue, because of true love? We step in, or his or her caregiver does: blue no more. . . . Can we stop war? We can sure as heck slow it down! Suddenly the soldiers on both sides start fucking. Or, at low doses, feeling super-fond."

It's funny, right? When we read it, we laugh. But we also think of classic sci-fi dystopias, where emotions are manipulated so the line between real and manufactured feelings blurs.

Yet, Americans are already enthusiastic brain tweakers. According to the Centers for Disease Control and Prevention, about one in ten takes an antidepressant. Students routinely, and successfully, use Ritalin to improve their mental focus for studying. Shift workers, pilots, truck drivers, and, yes, scientists use modafinil, a stay-awake drug, so they can keep working. We're not even counting the numbers of weed smokers, cocaine users, bourbon drinkers, nicotine addicts, or caffeine freaks. We use all these drugs for any number of reasons, including to cope with love (whether we want it and don't have it, have it but don't want it, or are trying to get over once having had it and losing it).

Many already use substances, especially alcohol, to enhance social experiences. Young ravers and partyers use MDMA, better known as ecstasy, because it gives them a feeling of belonging and friendship with their fellow glow-stick wavers. It does so, in part, because it's stimulating the release of oxytocin and dopamine.

In 2009, Larry wrote an essay for the journal *Nature* in which he speculated, as we have argued here, that love is a property that emerges from a series of chemical reactions in our brains. The essay prompted *New York Times* columnist John Tierney to speculate about a possible "love vaccine" against the emotion for use by the newly divorced or somebody in love with a person who could not, or would not, return that love. The column was republished around the world. Then Larry received this letter from a man in Nairobi: "I now humbly request that you direct me how I can get the vaccine for future prevention. It's to my hope you'll direct me, and if possible send me the vaccine." He was so interested, he wrote a follow-up letter: "If there is a cure drug on the same, I would like a few doses."

Wouldn't we all? Who hasn't been captured by an impossible love or felt the hollow ache of love gone bad? Who wouldn't want a shot of something to take all that pain away?

The poor fellow in Nairobi and the people who buy Liquid Trust *do* want to manipulate their and others' emotions. That's been the goal of witches, love-potion conjurers, and aphrodisiac-snake-oil salesmen for thousands of years. We feel one way, or somebody else feels one way, and we wish we, or they, felt another way.

In India, where oxytocin is already frequently used to force cows to give more milk, and even to enhance the look of vegetables, the press follows social-bonding experiments closely, partly because marriages are often arranged, rather than being love matches. If a drug could induce passion after the fact, many couples would use it.

If it's possible to harness the mechanisms we've discussed in this book—and we think it will be—they'll be harnessed. But that won't make love any less real.

The flower, jewelry, champagne, and perfume industries exist on our belief that such manipulation isn't only possible but desirable. As a society, we've already decided it's OK to use drugs to affect our personalities; there's no point in denying this. We don't generally feel these induced alterations in our behavior make our emotional experiences somehow less "real" than before. We're always being manipulated or manipulating ourselves and others. Love sparked by a drug wouldn't be any different from love sparked by martinis, clever banter, or good sex. The emotion is the same. If one accepts Larry's concept of love, then it doesn't matter what triggers the brain mechanisms. The activation of those circuits is what matters. However we got there, we'll act as if we made a choice. Love induced by a drug would still be love, real and true—or at least as real and true as any other.

We can know exactly how love, desire, and gender work in our brains, yet we'll still invent meaning to go along with that knowledge. We'll still celebrate the feelings and the thrills, as well as lament the sadness.

Now, though, we have a chance to be more willful, and more conscious of what we're doing. We have the opportunity to end uninformed prejudice, to appreciate the power of the love mechanism, and to try—albeit often futilely—to guard against heedlessness. Like those who don't believe in a god, or in a life after death, yet who construct ethical lives and find meaning despite their conviction that no supreme being is waiting to reward them, we'll make myths about meeting the person we love, about seeing our child's face for the first time, about the turbulent delight of our own sexual awakenings. Surely there will be deliberate, conscious acts of evil, too; it won't all be good.

But, says Wolpe, "let's say Larry is a hundred percent correct" in his hypotheses about love and bonding and desire. "Let's say I believe it to the bottom of my toes. Well, so what? How should that change how I behave? How I feel about my wife and children?"

It won't, and it doesn't have to. Wolpe and his wife and children have a

REWRITING THE STORY OF LOVE? · 263

narrative they've created around their shared experience of familial love. Even Larry, the guy who spends his days thinking about love and bonding in terms of biochemical reactions in localized circuits of the brain, experiences love for his wife and children as a passion undiminished by his reductionist perspective.

We do this every time we see a movie. Who can watch *Pride of the Yankees*, the story of Lou Gehrig, and not cry? We know the director, actors, and screenwriter are manipulating us, but we revel in our emotions anyway. We need the story for the lessons it teaches us about bravery, dignity, and, yes, love. Just so, once our appetite is engaged, we'll create reasons for our rapid heartbeats and the tingling in our loins. Back in Minnesota, Susan will still flirt. Even if she becomes conscious of it, and can dissect the neurochemistry involved, she'll still tell herself another tale for why she's doing it.

Of course, love will still go bad, just as it always has, and we will tell stories about that, too. Maybe, though, the new science can help smooth some of the most dangerous, most pathological manifestations of tragic love.

It's true we'll have to rethink some of the suppositions we've lived with for generations. But many who have long been denied full membership in the human social family, either because of biology or prejudice, will be able to demand their just due. We may also have to think again about how to navigate a human-relationship landscape that we know is so heavily influenced by the unconscious actions of chemicals on circuits. We may have to ask ourselves if what's right is always what's natural. If not, we'll have to decide when and how to make the distinction.

The new social neuroscience not only challenges us to think about these questions, but can also help provide solutions by informing the creation of new, even sturdier pillars we can use to support the culture of human love. If we're going to avoid the darker possibilities the future will bring, we'll have to be very thoughtful about what we tell ourselves. If we are, love won't ever fall off its pedestal.

ACKNOWLEDGMENTS

We're grateful to those whose cooperation made this book possible. We imposed on our interview subjects, sometimes prying into deeply personal matters, and they responded with the gifts of their time and forbearance. We are especially appreciative of those scientists who willingly engaged in spirited, sometimes risky, conversation about the implications of their work.

David Moldawer strongly believed in the idea, and Jillian Gray displayed a patient hand. Thanks also to Michelle Tessler, Susan Heard for her hard work and suggestions, and Alex Heard for his usual wise guidance. Brian gives a nod to Larry, a more pleasant and eager collaborator than anybody has a right to expect. Larry's wife, Anne, not only listens to his ideas about sex and relationships, she loves him anyway—which is saying something—and his children are constant reminders of why the story of love is so important.

BIBLIOGRAPHY

BUILDING A SEXUAL BRAIN

Amateau, S., and M. McCarthy. "Induction of PGE2 by Estradiol Mediates Developmental Masculinization of Sex Behavior." *Nature Neuroscience*, June 7, 2004.

Arnold, A. "The Organizational-Activational Hypothesis as the Foundation for a Unified Theory of Sexual Differentiation of All Mammalian Tissues." *Hormones and Behavior*, May 2009.

Auyeung, B., et al. "Fetal Testosterone Predicts Sexually Differentiated Childhood Behavior in Girls and in Boys." *Psychological Science*, February 2009.

Bao, A-M., and D. Swaab."Sex Differences in the Brain, Behavior, and Neuropsychiatric Disorders." *The Neuroscientist*, October 2010.

Berenbaum, S., et al. "Early Androgens Are Related to Childhood Sex-Typed Toy Preferences." *Psychological Science*, March 1992.

Berglund, H., et al. "Brain Response to Putative Pheromones in Lesbian Women." *Proceedings of the National Academy of Sciences*," May 23, 2006.

Bleier, R. "Why Does a Pseudohermaphrodite Want to Be a Man?" *New England Journal of Medicine*, October 11, 1979.

Bradley, S., et al. "Experiment of Nurture: Ablatio Penis at 2 Months, Sex Reassignment at 7 Months, and a Psychosexual Follow-Up in Young Adulthood." *Pediatrics*, July 1998.

Brooks, C. "Some Perversion of the Sexual Instinct." *Journal of the National Medical Association*, January–March 1919.

Capel, B., and D. Coveney. "Frank Lillie's Freemartin: Illuminating the Pathway to 21st Century Reproductive Endocrinology." *Journal of Experimental Zoology* 301 (2004).

Chura, L., et al. "Organizational Effects of Fetal Testosterone on Human Corpus Callosum Size and Asymmetry." *Psychoneuroendocrinology*, January 2010.

Ciumas, C., et al. "High Fetal Testosterone and Sexually Dimorphic Cerebral Networks in Females." *Cerebral Cortex*, May 2009.

Collaer, M., et al. "Motor Development in Individuals with Congenital Adrenal Hyperplasia: Strength, Targeting, and Fine Motor Skill."*Psychoneuroendocrinology*, February 2009.

Diamond, M. "Developmental, Sexual and Reproductive Neuroendocrinology: Historical, Clinical and Ethical Considerations." *Frontiers in Neuroendocrinology*, February 18, 2011.

Diamond, M. "Pediatric Management of Ambiguous and Traumatized Genitalia." *Journal of Urology*, September 1999.

Diamond, M. Interview with the authors, April 6, 2011.

Diamond, M., and K. Sigmundson. "Sex Reassignment at Birth: A Long Term Review and Clinical Implications." *Archives of Pediatrics and Adolescent Medicine*, March 1997.

Domurat Dreger, A. "'Ambiguous Sex'—or Ambivalent Medicine? Ethical Issues in the Treatment of Intersexuality." *Hastings Center Report* 28, no. 3 (1998).

Durante, M., et al. "Ovulation, Female Competition, and Product Choice: Hormonal Influences on Consumer Behavior." *Journal of Consumer Research*, April 2011.

Eckert, C. "Intervening in Intersexualization: The Clinic and the Colony." Dissertation. Proefschrift Universiteit Utrecht, 2010.

Ehrhardt, A., and H. Meyer-Bahlburg. "Effects of Prenatal Sex Hormones on Gender-Related Behavior." *Science,* March 20, 1981.

Garcia-Falgueras, A., and D. Swaab. "A Sex Difference in the Hypothalamic Uncinate Nucleus: Relationship to Gender Identity." *Brain*, November 2, 2008.

Gizewski, E., et al. "Specific Cerebral Activation Due to Visual Erotic Stimuli in Male-to-Female Transsexuals Compared with Male and Female Controls." *Journal of Sexual Medicine*, February 2009.

Glickman, S., et al. "Mammalian Sexual Differentiation: Lessons from the Spotted Hyena." *Trends in Endocrinology and Metabolism*, November 2006.

Goldstein, J., et al. "Normal Sexual Dimorphism of the Adult Human Brain Assessed by *In Vivo* Magnetic Resonance Imaging." *Cerebral Cortex*, June 2001.

Gooren, L. "Care of Transsexual Persons." *New England Journal of Medicine*, March 31, 2011.

Gorski, R. Interview with the authors, May 2, 2011.

Gorski, R. "Hypothalamic Imprinting by Gonadal Steroid Hormones." *Advances in Experimental Medicine and Biology*, 2002.

Gorski, R. "Sexual Dimorphisms of the Brain." *Journal of Animal Science* 61, supp. 3 (1985).

Gorski, R., et al. "Evidence for a Morphological Sex Difference Within the Medial Preoptic Area of the Rat Brain." *Brain Research*, June 16, 1978.

Greer, G. *The Whole Woman*. New York: Anchor Books, 2000.

Guerrero, L. Interview with the authors, March 8, 2011.

Hamann, S., et al. "Men and Women Differ in Amygdala Response to Visual Sexual Stimuli." *Nature Neuroscience*, April 2004.

Hasbro. http://www.hasbro.com/babyalive/en_us/shop/browse.cfm.

Hassett, J., et al. "Social Segregation in Male, but Not Female Yearling Rhesus Macaques." *American Journal of Primatology*, October 13, 2009.

Hassett, J., et al. "Sex Differences in Rhesus Monkey Toy Preferences Parallel Those of Children." *Hormones and Behavior*, August 2008.

Herman, R., et al. "Sex Differences in Interest in Infants in Juvenile Rhesus Monkeys: Relationship to Prenatal Androgen." *Hormones and* Behavior, May 2003.

Hines, M. "Prenatal Endocrine Influences on Sexual Orientation and on Sexually Differentiated Childhood Behavior." *Frontiers in Neuroendocrinology*, February 2001.

Hines, M., and G. Alexander. "Commentary: Monkeys, Girls, Boys and Toys: A Confirmation Comment on 'Sex Differences in Toy Preferences: Striking Parallels Between Monkeys and Humans.'" *Hormones and Behavior*, August 2008.

Imperato-McGinley, J., et al. "Androgens and the Evolution of Male-Gender Identity Among Male Pseudohermaphrodites with 5a-Reductase Deficiency." *New England Journal of Medicine*, May 31, 1979.

Imperato-McGinley, J., et al. "Steroid 5-alpha-reductase Deficiency in Man: An Inherited Form of Male Pseudohermaphroditism." *Science*, December 27, 1974.

Jacobson, C., et al. "The Influence of Gonadectomy, Androgen Exposure, or a Gonadal Graft in the Neonatal Rat on the Volume of the Sexually Dimorphic Nucleus of the Preoptic Area." *Journal of Neuroscience*, October 1, 1981.

Kahlenberg, S., and R. Wrangham. "Sex differences in chimpanzees' use of sticks as play objects resemble those of children." *Current Biology*, December 21, 2010.

Kimchi, T., et al. "A Functional Circuit Underlying Male Sexual Behavior in the Female Mouse Brain." *Nature*, August 30, 2007.

Kruijver, F., et al. "Male-to-Female Transsexuals Have Female Neuron Numbers in a Limbic Nucleus." *Journal of Endocrinology and Metabolism* 85, no. 5 (2000).

LeVay, S. "From Mice to Men: Biological Factors in the Development of Sexuality." *Frontiers in Neuroendocrinology*, February 2011.

Lillie, F. "Sex-Determination and Sex-Differentiation in Mammals." *Zoology*, July 1917.

Lillie, F. "The Theory of the Free-Martin," *Science*, April 28, 1916.

Marentette, J., et al. "Multiple Male Reproductive Morphs in the Invasive Round Goby (*Appollonia melanostoma*)." *Journal of Great Lakes Research*, June 2009.

McCarthy, M., et al. "New Tricks by an Old Dogma: Mechanisms of the Organizational/Activational Hypothesis of Steroid-Mediated Sexual Differentiation of Brain and Behavior." *Hormones and Behavior* 55 (2009).

Meyer-Bahlburg, H. "Gender Identity Outcome in Female-Raised 46,XY Persons with Penile Agenesis, Cloacal Exstrophy of the Bladder, or Penile Ablation." *Archives of Sexual Behavior*, August 2005.

Meyer-Bahlburg, H., et al. "Sexual Orientation in Women with Classical or Non-Classical Congenital Adrenal Hyperplasia as a Function of Degree of Prenatal Androgen Excess." *Archives of Sexual Behavior*, February 2008.

Money, J. "Ablatio Penis: Normal Male Infant Sex-Reassigned as a Girl." *Archives of Sexual Behavior*, January 1975.

Money, J., and J. Dalery. "Iatrogenic Homosexuality: Gender Identity in Seven 46,XX Chromosomal Females with Hyperadrenocortical Hermaphroditism Born with a Penis, Three Reared as Boys, Four Reared as Girls." *Journal of Homosexuality*, 1976.

Ngun, T., et al. "The Genetics of Sex Differences in Brain and Behavior." *Frontiers in Neuroendocrinology*, October 2010.

Ostrer, H., et al. "Mutations in MAP3K1 Cause 46,XY Disorders of Sex Development and Implicate a Common Signal Transduction Pathway in Human Testis Determination." *American Journal of Human Genetics*, December 2010.

Palanza, P., et al. "Effects of Developmental Exposure to Bisphenol A on Brain and Behavior in Mice." *Environmental Research*, October 2008.

Park, D., et al. "Male-like Sexual Behavior of Female Mouse Lacking Fucose Mutarotase." *BMC Genetics*, July 7, 2010.

Perkins, A., and C. Roselli. "The Ram as a Model for Behavioral Neuroendocrinology." *Hormones and Behavior*, June 2007.

Perkins, A., and J. A. Fitzgerald. "Luteinizing Hormones, Testosterone, and Behavioral Response of Male-Oriented Rams to Estrous Ewes and Rams." *Journal of Animal Science* 70 (1992).

Peterson, R., et al. "Male Pseudohermaphroditism Due to Steroid 5-alpha-reductase Deficiency." *American Journal of Medicine*, February 1977.

"Prenatal Shaping of Behavior." *British Medical Journal*, April 25, 1963.

Phoenix, C., et al. "Organizing Action of Prenatally Administered Testosterone Propionate on the Tissues Mediating Mating Behavior in the Female Guinea Pig." *Endocrinology*, September 1, 1959.

Renn, S., et al. "Fish and Chips: Functional Genomics of Social Plasticity in an African Cichlid Fish." *Journal of Experimental Biology*, September 2008.

Resko, A., et al. "Endocrine Correlates of Partner Preference Behavior in Rams." *Biology of Reproduction*, July 1, 1996.

Rosahn, P., and H. Greene. "The Influence of Intrauterine Factors on the Fetal Weight of Rabbits." *Journal of Experimental Medicine*, May 31, 1936.

Roselli, C. Interview with the authors, April 27, 2011.

Roselli, C., and F. Stormshak. "Prenatal Programming of Sexual Partner Preference: The Ram Model." *Journal of Neuroendocrinology*, March 2009.

Roselli, C., et al. "The Development of Male-Oriented Behavior in Rams." *Frontiers in Neuroendocrinology*, January 2011.

Roselli, C., et al. "The Ovine Sexually Dimorphic Nucleus of the Medial Preoptic Area Is Organized Prenatally by Testosterone." *Endocrinology*, May 2007.

Rupp, H., and K. Wallen. "Sex Differences in Viewing Sexual Stimuli: An Eye-Tracking Study in Men and Women." *Hormones and Behavior*, April 2007.

Ruta, V., et al. "A Dimorphic Pheromone Circuit in Drosophila from Sensory Input to Descending Output." *Nature*, December 2, 2010.

Saenger, P., et al. "Prepubertal Diagnosis of Steroid 5-alpha-reductase Deficiency." *Journal of Clinical Endocrinology and Metabolism*, April 1978.

Savic, I., and S. Arver. "Sex Dimorphism of the Brain in Male-to-Female Transsexuals." *Cerebral Cortex*, April 5, 2011.

Savic I., et al. "Male-to-Female Transsexuals Show Sex-Atypical Hypothalamus Activation When Smelling Odorous Steroids." *Cerebral Cortex*, August 2008.

Savic, I., et al. "PET and MRI Show Differences in Cerebral Asymmetry and Functional Connectivity Between Homo- and Heterosexual Subjects." *Proceedings of the National Academy of Sciences*, June 16, 2008.

Savic, I., et al. "Brain Response to Putative Pheromones in Homosexual Men." *Proceedings of the National Academy of Sciences*, May 17, 2005.

Schwartz, J. "Of Gay Sheep, Modern Science and Bad Publicity." *New York Times*, January 25, 2007.

Scott, H., et al. "Steroidogenesis in the Fetal Testis and Its Susceptibility to Disruption by Exogenous Compounds." *Endocrine Reviews*, November 3, 2009.

Sommer, V., and P. Vasey. *Homosexual Behavior in Animals: An Evolutionary Perspective*. London: Cambridge University Press, 2006.

Stowers, L., et al. "Loss of Sex Discrimination and Male-Male Aggression in Mice Deficient for TRP." *Science*, February 22, 2002.

Swaab, D. Interview with the authors, March 28, 2011.

Swaab, D. "Sexual Orientation and Its Basis in Brain Structure and Function." *Proceedings of the National Academy of Sciences*, July 29, 2008.

Swaab, D., and M. Hofman. "An Enlarged Suprachiasmatic Nucleus in Homosexual Men." *Brain Research*, December 24, 1990.

Swan, S. Interview with the authors, November 26, 2010.

"The Sexes: Biological Imperatives." *Time*, January 8, 1973.

Urological Sciences Research Foundation, http://www.usrf.org/news/010308-guevedoces.html.

Vom Saal, F. "Sexual Differentiation in Litter-Bearing Mammals: Influence of Sex of Adjacent Fetuses in Utero." *Journal of Animal Science*, July 1989.

Vom Saal, F., et al. "Chapel Hill Bisphenol A Expert Panel Consensus Statement: Integration of Mechanisms, Effects in Animals and Potential to Impact Human Health at Current Levels of Exposure." *Reproductive Toxicology*, August–September 2007.

Vom Saal, F., et al. "Paradoxical Effects of Maternal Stress on Fetal Steroids and Postnatal Reproductive Traits in Female Mice from Different Intrauterine Positions." *Biology of Reproduction*, November 1990.

Wallen, K. "The Organizational Hypothesis: Reflections on the 50th Anniversary of the Publication of Phoenix, Goy, Gerall, and Young (1959)." *Hormones and Behavior*, May 2009.

Wallen, K. "Sex and Context: Hormones and Primate Sexual Motivation." *Hormones and Behavior*, September 2001.

Wallen, K., and J. Hassett. "Sexual Differentiation of Behavior in Monkeys: Role of Prenatal Hormones." *Journal of Neuroendocrinology*, March 2009.

Wallen, K., and H. Rupp. "Women's Interest in Visual Sexual Stimuli Varies with Menstrual Cycle Phase at First Exposure and Predicts Later Interest." *Hormones and Behavior*, February 2010.

Whitam, F., et al. "Homosexual Orientation in Twins: A Report on 61 Pairs and Three Triplet Sets." *Archives of Sexual Behavior*, June 1993.

Wilson, J. "Androgens, Androgen Receptors, and Male Gender Role Behavior." *Hormones and Behavior*, September 2001.

Women Studies Department, University of Wisconsin. http://womenstudies.wisc.edu/ruthbleier-scholarship.htm.

Woodson, J., et al. "Sexual Experience Interacts with Steroid Exposure to Shape the Partner Preferences of Rats." *Hormones and Behavior*, September 2002.

Young, L., and D. Crews. "Comparative Neuroendocrinology of Steroid Receptor Gene Expression and Regulation: Relationship to Physiology and Behavior." *Trends in Endocrinology and Metabolism*, September/October 1995.

Young, L., and Z. Wang. "The Neurobiology of Pair Bonding." *Nature Neuroscience*, October 2004.

Zhou, J., et al. "A Sex Difference in the Human Brain and Its Relation to Transsexuality." *International Journal of Transgenderism*, September 1997.

THE CHEMISTRY OF DESIRE

Baird, A., et al. "Neurological Control of Human Sexual Behavior: Insights from Lesion Studies." *Journal of Neurology, Neurosurgery and Psychiatry*, December 22, 2006.

Brown, S., et al. "The Menstrual Cycle and Sexual Behavior: Relationship to Eating, Exercise, Sleep and Health Patterns." *Women and Health*, May 20, 2009.

Burnham, T. "High-Testosterone Men Reject Low Ultimatum Game Offers." *Proceedings of the Royal Society B*, July 5, 2007.

Davidson, J., and G. Bloch. "Neuroendocrine Aspects of Male Reproduction." *Biology of Reproduction* 1 (1969).

Desjardins, J., et al. "Female Genomic Response to Mate Information." *Proceedings of the National Academy of Sciences*, December 7, 2010.

Durante, K. Interview with the authors, April 19, 2011.

Durante, K., and N. Li. "Oestradiol Level and Opportunistic Mating in Women." *Biology Letters,* January 13, 2009.

Durante, K., et al. "Changes in Women's Choice of Dress Across the Ovulatory Cycle: Naturalistic and Laboratory Task-Based Evidence." *Personality and Social Psychology Bulletin,* August 21, 2008.

Everitt, B. "Sexual Motivation: A Neural and Behavioral Analysis of the Mechanisms Underlying Appetitive and Copulatory Response of Male Rats." *Neuroscience and Biobehavioral Reviews* 14 (1990).

Fessler, D. "No Time to Eat: An Adaptationist Account of Periovulatory Behavioral Changes." *Quarterly Review of Biology,* March 2003.

Fleischman, D., and D. Fessler. "Differences in Dietary Intake as a Function of Sexual Activity and Hormonal Contraception." *Evolutionary Psychology* 5 (2007).

Gangestad, S., and R. Thornhill. "Human Oestrus." *Proceedings of the Royal Society B,* February 5, 2008.

Gangestad, S., et al. "Changes in Women's Mate Preferences Across the Ovulatory Cycle." *Journal of Personality and Social Psychology,* January 2007.

Gangestad, S., et al. "Women's Preferences for Male Behavioral Displays Change Across the Menstrual Cycle." *Psychological Science,* March 15, 2004.

Gangestad, S., et al. "Changes in Women's Sexual Interests and Their Partners' Mate-Retention Tactics Across the Menstrual Cycle: Evidence for Shifting Conflicts of Interest." *Proceedings of Biological Sciences,* May 7, 2002.

Goldstein, J., et al. "Hormonal Cycle Modulates Arousal Circuitry in Women Using Functional Magnetic Resonance Imaging." *Journal of Neuroscience,* October 5, 2005.

Griskevicius, V., et al. "Aggress to Impress: Hostility As an Evolved Context-Dependent Strategy." *Journal of Personality and Social Psychology* 96, no. 5 (2009).

Harris, G., and R. Michael. "The Activation of Sexual Behavior by Hypothalamic Implants of Oestrogen." *Journal of Physiology* 171, no. 2 (1964).

Haselton, M., and S. Gangstad. "Conditional Expression of Women's Desires and Men's Mate Guarding Across the Ovulatory Cycle." *Hormones and Behavior,* January 3, 2006.

Haselton, M., and K. Gildersleeve. "Can Men Detect Ovulation?" *Current Directions in Psychological Science,* April 2011.

Haselton, M., et al. "Ovulatory Shifts in Human Female Ornamentation: Near Ovulation, Women Dress to Impress." *Hormones and Behavior* 51 (2007).

Hill, S., and D. Buss. "The Mere Presence of Opposite-Sex Others on Judgments of Sexual and Romantic Desirability: Opposite Effects for Men and Women." *Personality and Social Psychology Bulletin,* February 26, 2008.

Hill, S., and K. Durante. "Courtship, Competition, and the Pursuit of Attractiveness: Mating Goals Facilitate Health-Related Risk Taking and Strategic Risk

Suppression in Women." *Personality and Social Psychology Bulletin*, January 20, 2001.

Hill, S., and K. Durante. "Do Women Feel Worse to Look Their Best? Testing the Relationship Between Self-Esteem and Fertility Status Across the Menstrual Cycle." *Personality and Social Psychology Bulletin*, September 17, 2009.

Hill, S., and M. Ryan. "The Role of Model Female Quality in the Mate Choice Copying Behavior of Sailfin Mollies." *Biology Letters*, December 19, 2005.

Hull, E., and J. Dominguez. "Sexual Behavior in Male Rodents." *Hormones and Behavior*, April 19, 2007.

Kimchi, T., et al. "A Functional Circuit Underlying Male Sexual Behavior in the Female Mouse Brain." *Nature*, August 30, 2007.

Kruger, D. "When Men Are Scarce, Good Men Are Even Harder to Find: Life History, the Sex Ratio, and the Proportion of Men Married." *Journal of Social, Evolutionary, and Cultural Psychology* 3 (2009).

Levi, M., et al. "Deal or No Deal: Hormones and the Mergers and Acquisitions Game." *Management Science*, September 2010.

McClintock, M., et al. "Human Body Scents: Conscious Perceptions and Biological Effects." *Chemical Senses* 30, supp. 1 (2005).

Michael, R., and P. Scott. "The Activation of Sexual Behaviour in Cats by the Subcutaneous Administration of Oestrogen." *Journal of Physiology* 171, no. 2 (1964).

Miller, G. Interview with the authors, May 9, 2011.

Miller, G., et al. "Ovulatory Cycle Effects on Tip Earnings by Lap Dancers: Economic Evidence for Human Estrus?" *Evolution and Human Behavior* 28 (2007).

Miller, S., and J. Maner. "Ovulation as a male mating prime: subtle signs of women's fertility influence men's mating cognition and behavior." *Journal of Personality and Social Psychology,* February 2011.

Miller, S., and J. Maner. "Scent of a woman: men's testosterone responses to olfactory ovulation cues." *Psychological Science,* February 2010.

Paris, C., et al. "A Possible Mechanism for the Induction of Lordosis by Reserpine in Spayed Rats." *Biology of Reproduction*, February 4, 1971.

Pfaff, D. "Autoradiographic Localization of Radioactivity in Rat Brain After Injection of Tritiated Sex Hormones." *Science*, September 27, 1968.

Pfaff, D., et al. "Reverse Engineering the Lordosis Behavior Circuit." *Hormones and Behavior*, April 2008.

Pfaus, G. "Pathways of Sexual Desire." *Journal of Sexual Medicine*, June 2006.

Pfaus, G., and M. Scepkowski. "The Biologic Basis for Libido." *Current Sexual Health Reports*, February 2005.

Pillsworth, E., et al. "Kin Affiliation Across the Ovulatory Cycle: Females Avoid Fathers When Fertile." *Psychological Science*, November 24, 2010.

Prudom, S., et al. "Exposure to Infant Scent Lowers Serum Testosterone in Father Common Marmosets." *Biology Letters*, August 26, 2008.

Rupp, H. Interview with the authors, May 10, 2011.

Rupp, H., and K. Wallen. "Relationship Between Testosterone and Interest in Sexual Stimuli: The Effect of Experience." *Hormones and Behavior*, August 10, 2007.

Rupp, H., et al. "Neural Activation in the Orbitofrontal Cortex in Response to Male Faces Increases During the Follicular Phase." *Hormones and Behavior*, June 2009.

Shille, V., et al. "Follicular Function in the Domestic Cat as Determined by Estradiol-17ß Concentrations in Plasma: Relation to Estrous Behavior and Cornification of Exfoliated Vaginal Epithelium." *Biology of Reproduction* 21 (1979).

Slatcher, R., et al. "Testosterone and Self-Reported Dominance Interact to Influence Human Mating Behavior." *Social Psychological and Personality Science*, February 28, 2011.

Sundie, J., et al. "Peacocks and Porsches and Thorstein Veblen: Conspicuous Consumption as a Sexual Signaling System." *Journal of Personality and Social Psychology*, November 1, 2010.

Takahashi, L. "Hormonal Regulation of Sociosexual Behavior in Female Mammals." *Neuroscience and Behavioral Reviews*, April 1990.

Wiesner, B., and L. Mirskaia. "On the Endocrine Basis of Mating in the Mouse." *Experimental Physiology*, October 9, 1930.

Zhu, X., et al. "Brain Activation Evoked by Erotic Films Varies with Different Menstrual Phases: An fMRI Study." *Behavioral Brain Research*, January 20, 2010.

Ziegler, T., et al. "Neuroendocrine Response to Female Ovulatory Odors Depends Upon Social Condition in Male Common Marmosets." *Hormones and Behavior*, January 2005.

THE POWER OF APPETITE

Abelard, P. *Historia Calamitatum*. Project Gutenberg. http://www.gutenberg.org/ebooks/14268.

Abelard, P., and Heloise. *Letters of Abelard and Heloise*. Project Gutenberg. http://gutenberg.org/ebooks/35977.

Alighieri, D. *The Divine Comedy*. Translated by John Ciardi. New York: W. W. Norton, 1977.

Ariely, D., and G. Loewenstein. "The Heat of the Moment: The Effect of Sexual Arousal on Sexual Decision Making." *Journal of Behavioral Decision Making*, July 26, 2005.

Associated Press, "Exec Admits Stealing from Charity for S&M Bill." March 28, 2006.

Balfour, M., et al. "Sexual Behavior and Sex-Associated Environmental Cues Activate the Mesolimbic System in Male Rats." *Neuropsychopharmacology*, December 23, 2003.

Barfield, R., and B. Sachs. "Sexual Behavior: Stimulation by Painful Electric Shock to Skin of Male Rats." *Science*, July 26, 1968.

Bishop, M., et al. "Intracranial Self-Stimulation in Man." *Science*, April 26, 1963.

Bullough, V., and J. Brundage, eds. *Handbook of Medieval Sexuality*. New York: Garland, 2000.

Caggiula, A., and B. Hoebel. "'Copulation-Reward Site' in the Posterior Hypothalamus." *Science*, May 9, 1966.

Childs, E., and H. de Wit. "Amphetamine-induced Place Preference in Humans." *Biological Psychiatry*, May 15, 2009.

Coolen, L. "Activation of mu-Opioid Receptors in the Medial Preoptic Area Following Copulation in Male Rats." *Neuroscience*, February 18, 2004.

Coria-Avila, G., et al. "Olfactory Conditioned Partner Preference in the Female Rat." *Behavioral Neuroscience*, vol. 119, no. 3, 2005.

"Dr. Robert G. Heath." *New York Times*, September 27, 1999.

Dominguez, J., and E. Hull. "Dopamine, the Medial Preoptic Area, and Male Sexual Behavior." *Physiology and Behavior*, October 15, 2005.

Dreher, J-C., et al. "Menstrual Cycle Phase Modulates Reward-Related Neural Function in Women." *Proceedings of the National Academy of Sciences*, February 13, 2007.

Eagle, D., et al. "Contrasting Roles for Dopamine D1 and D2 Receptor Subtypes in the Dorsomedial Striatum but Not the Nucleus Accumbens Core During Behavioral Inhibition in the Stop-Signal Task in Rats." *Journal of Neuroscience*, May 18, 2011.

Elliott, V. "Patients Returning to Cosmetic Surgery as Recession Loosens Grip." *AMA News*, February 28, 2011.

Everitt, B. "Sexual Motivation: A Neural and Behavioural Analysis of the Mechanisms Underlying Appetitive and Copulatory Responses of Male Rats." *Neuroscience and Biobehavioural Reviews* 14 (1990): 217–32.

Flagel, S., et al. "A Selective Role for Dopamine in Stimulus-Reward Learning." *Nature*, January 6, 2011.

Frohmader, K., et al. "Methamphetamine Acts on Subpopulations of Neurons Regulating Sexual Behavior in Male Rats." *Neuroscience*, March 31, 2010.

Frohmader, K., et al. "Mixing Pleasures: Review of the Effects of Drugs on Sex Behavior in Humans and Animal Models." *Hormones and Behavior*, December 31, 2009.

Frohman, E., et al. "Acquired Sexual Paraphilia in Patients with Multiple Sclerosis." *Archives of Neurology* 59 (2002): 1006–10.

Groman, S., et al. "Dorsal Striatal D2-Like Receptor Availability Covaries with Sensitivity to Positive Reinforcement During Discrimination Learning." *Journal of Neuroscience*, May 18, 2011.

Hamann, S., et al. "Men and Women Differ in Amygdala Response to Visual Sexual Stimuli." *Nature Neuroscience*, April 2004.

Hamburger-Bar, R., and H. Rigter. "Apomorphine: Facilitation of Sexual Behaviour in Female Rats." *European Journal of Pharmacology*, June–July 1975.

Hartmann, E. *The Philosophy of the Unconscious: Speculative Results According to the Inductive Method of Physical Science.* Edinburgh: Ballantyne, Hanson, and Company, 1884.

Heath, R. "Correlations Between Levels of Psychological Awareness and Physiological Activity in the Central Nervous System." *Psychosomatic Medicine* 17, no. 5 (1955).

Health, R., and E. Norman. "Electroshock Therapy by Stimulation of Discrete Cortical Sites with Small Electrodes." *Proceedings of the Society for Experimental Biology and Medicine. Society for Experimental Biology and Medicine*, December 1946.

Heath, R., et al. "Effects of Chemical Stimulation to Discrete Brain Areas." *American Journal of Psychiatry*, May 1, 1961.

Holder, M., and J. Mong. "Methamphetamine Enhances Paced Mating Behaviors and Neuroplasticity in the Medial Amygdala of Female Rats." *Hormones and Behavior*, April 24, 2010.

Hori, A., and T. Akimoto. "Four Cases of Sexual Perversions." *Kurume Medical Journal* 15, no. 3 (1968).

Hull, E., et al. "Hormone-Neurotransmitter Interactions in the Control of Sexual Behavior." *Behavioural Brain Research* 105 (1999): 105–16.

Katz, H. Interview with the authors, June 1, 2011.

Keiper, A. "The Age of Neuroelectronics." *New Atlantis*, Winter 2006.

Kelley, A., and K. Berridge. "The Neuroscience of Natural Rewards: Relevance to Addictive Drugs." *Journal of Neuroscience*, May 1, 2002.

Krafft-Ebing, R. *Psychpathia Sexualis.* http://www.archive.org/stream/sexualin stinctcon00krafuoft/sexualinstinctcon00krafuoft_djvu.txt.

Lee, S., et al. "Effect of Sertraline on Current-Source Distribution of the High-Beta Frequency Band: Analysis of Electroencephalography Under Audiovisual Erotic Stimuli in Healthy, Right-Handed Males." *Korean Journal of Urology*, August 18, 2010.

Liu, Y., et al. "Social Bonding Decreases the Rewarding Properties of Amphetamine Through a Dopamine D1 Receptor-Mediated Mechanism." *Journal of Neuroscience*, June 1, 2011.

Loewenstein, G. "Out of Control: Visceral Influences on Behavior." *Organizational Behavior and Human Decision Processes*, March 1996.

Loewenstein, G., et al. "The Effect of Sexual Arousal on Expectations of Sexual Forcefulness." *Journal of Research in Crime and Delinquency*, November 4, 1997.

Lorrain, D., et al. "Lateral Hypothalamic Serotonin Inhibits Nucleus Accumbens Dopamine: Implications for Sexual Satiety." *Journal of Neuroscience*, September 1, 1999.

Meisel, R., and A. Mullins. "Sexual Experience in Female Rodents: Cellular Mechanisms and Functional Consequences." *Brain Research*, December 18, 2006.

Mendez, M. "Hypersexuality After Right Pallidotomy for Parkinson's Disease." *Journal of Neuropsychiatry and Clinical Neuroscience*, February 2004.

Moan, C., and R. Heath. "Septal Stimulation for the Initiation of Heterosexual Behavior in a Homosexual Male." *Journal of Behavioral Therapy and Experimental Psychiatry* 3 (1972): 23–30.

Monroe, R., and R. Heath. "Effects of Lysergic Acid and Various Derivatives on Depth and Cortical Electrograms." *Journal of Neuropsychiatry*, November–December 1961.

O'Halloran, R., and P. Dietz. "Autoerotic Fatalities with Power Hydraulics." *Journal of Forensic Science*, March 1993.

Olds, J., and P. Milner. "Positive Reinforcement Produced by Electrical Stimulation of Septal Area and Other Regions of Rat Brain." *Journal of Comparative and Physiological Psychology*, December 1954.

Pfaus, J. Interview with the authors, June 8–9, 2011.

Pfaus, J. "Pathways of Sexual Desire." *Journal of Sexual Medicine*, June 6, 2009.

Pfaus, J. "Conditioning and Sexual Behavior: A Review." *Hormones and Behavior*, September 2001.

Pfaus, J., and L. Scepkowski. "The Biologic Basis for Libido." *Current Sexual Health Reports* 2 (2005): 95–100.

Pfaus, J., et al. "What Can Animal Models Tell Us About Human Sexual Response?" *Annual Review of Sex Research*, 2003.

Pitchers, K., et al. "Neuroplasticity in the Mesolimbic System Induced by Natural Reward and Subsequent Reward Abstinence." *Biological Psychiatry*, May 1, 2010.

"Playboy Enterprises Inc." Encyclopedia of Chicago. Chicago Historical Society, 2005.

Portenoy, R., et al. "Compulsive Thalamic Self-Stimulation: A Case with Metabolic, Electrophysiologic and Behavioral Correlates." *Pain*, December 27, 1986.

Rankin, R. "Judge Says Depression, Accident Led to Cocaine, Stripper Troubles." *Atlanta Journal-Constitution*, February 26, 2011.

Rupp, H. Interview with the authors, May 10, 2011.

Rupp, H. "The Role of the Anterior Cingulate Cortex in Women's Sexual Decision Making." *Neuroscience Letters*, January 2, 2009.

Scaletta, L., and E. Hull. "Systemic or Intracranial Apomorphine Increases Copulation in Long-Term Castrated Male Rats." *Pharmacology Biochemistry and Behavior*, November 1990.

Smith, A. "Cosmetic Surgery Market Stands Firm." CNNMoney.com, February 20, 2008.

Stuber, G., et al. "Excitatory Transmission from the Amygdala to Nucleus Accumbens Facilitates Reward Seeking." *Nature*, June 29, 2011.

Tabibnia, G., et al. "Different Forms of Self-Control Share a Neurocognitive Substrate." *Journal of Neuroscience*, March 30, 2011.

Takahashi, H., et al. "Dopamine D1 Receptors and Nonlinear Probability Weighting in Risky Choice." *Journal of Neuroscience*, December 8, 2010.

Taub, S. "Accountant Embezzled to Pay Dominatrix." CFO.com, May 4, 2006.

Tenk, C., et al. "Sexual Reward in Male Rats: Effects of Sexual Experience on Conditioned Place Preferences Associated with Ejaculation and Intromission." *Hormones and Behavior*, January 2009.

Thompson, R. "Biography of James Olds." http://www.nap.edu/readingroom.php?book=biomems&page=jolds.html.

Tomlinson, W. Interview with Robert Heath. http://www.archive.org/details/WallaceTomlinsonInterviewingRobertHeath_March51986.

Torpy, B., and S. Visser. "Seamy Allegations Just Don't Fit Courtly Life." *Atlanta Journal-Constitution*, October 10, 2010.

United States of America v. Jack T. Camp. Case Number 1:10-MJ-1415, October 4, 2010.

Well Blog. *New York Times.* http://well.blogs.nytimes.com/2010/03/09/sagging-interest-in-plastic-surgery.

Wen Wan, E., and N. Agrawal. "Carry-Over Effects on Decision-Making: A Construal Level Perspective." *Journal of Consumer Research*, June 2011.

Young, L., and Z. Wang. "The Neurobiology of Pair Bonding." *Nature Neuroscience*, September 26, 2004.

THE MOMMY CIRCUIT

Ahern, T., and L. Young. "The Impact of Early Life Family Structure on Adult Social Attachment, Alloparental Behavior, and the Neuropeptide Systems Regulating Affiliative Behaviors in the Monogamous Prairie Vole." *Frontiers in Behavioral Neuroscience*, August 27, 2009.

Ahern, T., et al. "Parental Division of Labor, Coordination, and the Effects of Family Structure on Parenting in Monogamous Prairie Voles." *Developmental Psychology*, March 2011.

Bartels, A., and S. Zeki. "The Neural Correlates of Maternal and Romantic Love." *Neuroimage*, March 2004.

Beckett, C., et al. "Do the Effects of Early Severe Deprivation on Cognition Persist into Early Adolescence? Findings from the English and Romanian Adoptees Study." *Child Development*, May–June 2006.

Bellow, S. *The Dean's December.* New York: Pocket Books, 1982.

Bos, K., et al. "Effects of Early Psychosocial Deprivation on the Development of Memory and Executive Function." *Frontiers in Behavioral Neuroscience*, September 1, 2009.

Bos, P., et al. "Acute Effects of Steroid Hormones and Neuropeptides on Human Social-Emotional Behavior: A Review of Single Administration Studies."*Frontiers in Neuroendocrinology*, January 21, 2011.

Broad, K., et al. "Mother-Infant Bonding and the Evolution of Mammalian Social Relationships." *Philosophical Transactions of the Royal Society B*, November 6, 2006.

Buchen, L. "In Their Nurture." *Nature*, September 9, 2010.

Buckley, J. Interview with the authors, August 10, 2011.

Cameron, N. "Maternal Influences on the Sexual Behavior and Reproductive Success of the Female Rat." *Hormones and Behavior*, June 2008.

Cameron, N., et al. "Maternal Programming of Sexual Behavior and Hypothalamic-Pituitary-Gonadal Function in the Female Rat." *PLoS One*, May 21, 2008.

Champagne, D., et al. "Maternal Care and Hippocampal Plasticity: Evidence for Experience-Dependent Structural Plasticity, Altered Synaptic Functioning, and Differential Responsiveness to Glucocorticoids and Stress." *Journal of Neuroscience*, June 4, 2008.

Champagne, F. Interview with the authors, March 24, 2011.

Champagne, F., and M. Meaney. "Transgenerational Effects of Social Environment on Variations in Maternal Care and Behavioral Response to Novelty." *Behavioral Neuroscience*, December 2007.

Champagne, F., et al. "Maternal Care Associated with Methylation of the Estrogen Receptor-α1b Promoter and Estrogen Receptor-α Expression in the Medial Pre-optic Area of Female Offspring." *Endocrinology*, March 2, 2006.

Champagne, F., et al. "Naturally Occurring Variations in Maternal Behavior in the Rat Are Associated with Differences in Estrogen-Inducible Central Oxytocin Receptors." *Proceedings of the National Academy of Sciences*, October 23, 2001.

Da Costa, A. "The Role of Oxytocin Release in the Paraventricular Nucleus in the Control of Maternal Behaviour in the Sheep." *Journal of Neuroendocrinology*, March 1996.

Da Costa, A., et al. "Face Pictures Reduce Behavioural, Autonomic, Endocrine and Neural Indices of Stress and Fear in Sheep." *Proceedings of The Royal Society B*, September 7, 2004.

Deardorff, J., et al. "Father Absence, BMI and Pubertal Timing in Girls: Differential Effects by Family Income and Ethnicity." *Journal of Adolescent Health*, September 17, 2010.

Devlin, A., et al. "Prenatal Exposure to Maternal Depressed Mood and the MTH-FRC677T Variant Affect SLC6A4 Methylation in Infants at Birth." *PLoS One*, August 16, 2010.

Febo, M., et al. "Functional Magnetic Resonance Imaging Shows Oxytocin Activates Brain Regions Associated with Mother-Pup Bonding During Suckling." *Journal of Neuroscience*, December 14, 2005.

Fleming, A., et al. "Father of Mothering: Jay S. Rosenblatt." *Hormones and Behavior*, January 14, 2009.

Francis, D., et al. "Naturally Occurring Differences in Maternal Care Are Associated with the Expression of Oxytocin and Vasopressin (V1a) Receptors: Gender Differences." *Journal of Neuroendocrinology*, May 2002.

Francis, D., et al. "Variations in Maternal Behaviour Are Associated with Differences in Oxytocin Receptor Levels in the Rat." *Journal of Neuroendocrinology*, December 2000.

George, E., et al. "Maternal Separation with Early Weaning: A Novel Mouse Model of Early Life Neglect." *BMC Neuroscience*, September 29, 2010.

Grady, D. "Cesarean Births Are at a High in U.S." *New York Times*, March 23, 2010.

Hamilton, B., and S. Ventura. "Fertility and Abortion Rates in the United States, 1960–2002." *International Journal of Andrology*, February 2006.

Heim, C., et al. "Effect of Childhood Trauma on Adult Depression and Neuroendocrine Function: Sex-Specific Moderation by CRH Receptor 1 Gene." *Frontiers in Behavioral Neuroscience*, November 6, 2009.

Heim, C., et al. "Lower CSF Oxytocin Concentrations in Women with a History of Childhood Abuse." *Molecular Psychiatry*, October 2009.

Heim, C., et al. "Pituitary-Adrenal and Autonomic Responses to Stress in Women After Sexual and Physical Abuse in Childhood." *Journal of the American Medical Association*, August 2, 2000.

Henshaw, S. "Unintended Pregnancy in the United States." *Family Planning Perspectives*, January–February 1998.

Kendrick, K., et al. "Sex Differences in the Influence of Mothers on the Sociosexual Preferences of Their Offspring." *Hormones and Behavior*, September 2001.

Kim, P., et al. "Breastfeeding, Brain Activation to Own Infant Cry, and Maternal Sensitivity." *Journal of Child Psychology and Psychiatry*, August 2011.

Kim, P., et al. "The Plasticity of Human Maternal Brain: Longitudinal Changes in Brain Anatomy During the Early Postpartum Period." *Behavioral Neuroscience*, October 2010.

Kinnally, E., et al. "Epigenetic Regulation of Serotonin Transporter Expression and Behavior in Infant Rhesus Macaques." *Genes, Brain and Behavior*, August 2010.

Kramer, J. "Against Nature." *New Yorker*, July 25, 2011.

Marshall, G. Interview with the authors, August 24, 2011.

Marshall, M. Interview with the authors, August 24, 2011.

Maselko, J., et al. "Mother's Affection at 8 Months Predicts Emotional Distress in Adulthood." *Journal of Epidemiology and Community Health*, July 2011.

Matthiesen, A. S., et al. "Postpartum Maternal Oxytocin Release by Newborns: Effects of Infant Hand Massage and Sucking." *Birth*, March 2001.

McGowan, P. O., et al. "Broad Epigenetic Signature of Maternal Care in the Brain of Adult Rats." *PLoS One*, February 28, 2011.

McGowan, P. O., et al. "Epigenetic Regulation of the Glucocorticoid Receptor in Human Brain Associates with Childhood Abuse." *Nature Neuroscience*, March 2009.

Meaney M. "Epigenetics and the Biological Definition of Gene X Environment Interactions." *Child Development*, January 2010.

Meinlschmidt, G., and C. Heim. "Sensitivity to Intranasal Oxytocin in Adult Men with Early Parental Separation." *Biological Psychiatry*, May 1, 2077.

Numan, M., and B. Woodside. "Maternity: Neural Mechanisms, Motivational Processes, and Physiological Adaptations." *Behavioral Neuroscience*, December 2010.

Numan, M., et al. "Medial Preoptic Area and Onset of Maternal Behavior in the Rat." *Journal of Comparative and Physiological Psychology*, February 1977.

Oberlander, T., et al. "Prenatal Exposure to Maternal Depression, Neonatal Methylation of Human Glucocorticoid Receptor Gene (NR3C1) and Infant Cortisol Stress Responses." *Epigenetics*, March–April 2008.

Olazabal, D., and L. Young. "Oxytocin Receptors in the Nucleus Accumbens Facilitate 'Spontaneous' Maternal Behavior in Adult Female Prairie Voles." *Neuroscience*, August 25, 2006.

Olazabal, D., and L. Young. "Species and Individual Differences in Juvenile Female Alloparental Care Are Associated with Oxytocin Receptor Density in the Striatum and the Lateral Septum." *Hormones and Behavior*, May 2006.

Pedersen, C., and A. Prange. "Induction of Maternal Behavior in Virgin Rats After Intracerebroventricular Administration of Oxytocin." *Proceedings of the National Academy of Sciences*, December 1979.

Porter, R., et al. "Induction of Maternal Behavior in Non-Parturient Adoptive Mares." *Physiology and Behavior*, September 2002.

Pruessner, J., et al. "Dopamine Release in Response to a Psychological Stress in Humans and Its Relationship to Early Life Maternal Care: A Positron Emission Tomography Study Using [^{11}C]Raclopride." *Journal of Neuroscience*, March 17, 2004.

Rosenblatt, J. "Nonhormonal Basis of Maternal Behavior in the Rat." *Science*, June 16, 1967.

Rosenblatt, J., et al. "Hormonal Basis During Pregnancy for the Onset of Maternal Behavior in the Rat." *Psychoneuroendocrinology* 13 (1988): 29–46.

Ross, H., and L. Young. "Oxytocin and the Neural Mechanisms Regulating Social Cognition and Affiliative Behavior." *Frontiers in Neuroendocrinology*, May 28, 2009.

Seltzer, L., et al. "Social Vocalizations Can Release Oxytocin in Humans." *Proceedings of The Royal Society B*, March 6, 2010.

Shahrokh, D., et al. "Oxytocin-Dopamine Interactions Mediate Variations in Maternal Behavior in the Rat." *Endocrinology*, March 12, 2010.

Shakespeare, W. *Macbeth*. New York: Signet Classic, 1963.

Strathearn, L. Interview with the authors, August 31, 2011.

Strathearn, L., et al. "Adult Attachment Predicts Maternal Brain and Oxytocin Response to Infant Cues." *Neuropsychopharmacology*, December 2009.

Strathearn, L., et al. "What's in a Smile? Maternal Brain Responses to Infant Cues." *Pediatrics*, July 2008.

Sullivan, R., et al. "Developing a Neurobehavioral Animal Model of Infant Attachment to an Abusive Caregiver." *Biological Psychiatry*, June 15, 2010.

Swain, J., et al. "Maternal Brain Response to Own Baby-Cry Is Affected by Cesarean Section Delivery." *Journal of Child Psychology and Psychiatry*, October 2008.

SweetnessInFlorida. http://forums.plentyoffish.com/13867208datingPostpage2.aspx.

Terkel, J., and J. Rosenblatt. "Maternal Behavior Induced by Maternal Blood Plasma Injected Into Virgin Rats." *Journal of Comparative and Physiological Psychology*, June 1968.

Wismer Fries, A., et al. "Early Experience in Humans Is Associated with Changes in Neuropeptides Critical for Regulating Social Behavior." *Proceedings of the National Academy of Sciences*, November 22, 2005.

BE MY BABY

Abelard, P. *Historia Calamitatum*. Project Gutenberg. http://gutenberg.org/ebooks/14268.

Abelard, P., and Heloise. *Letters of Abelard and Heloise*. Project Gutenberg. http://gutenberg.org/ebooks/35977.

Ackerman, J., et al. "Let's Get Serious: Communicating Commitment in Romantic Relationships." *Journal of Personality and Social Psychology*, June 2011.

Aragona, B., et al. "Nucleus Accumbens Dopamine Differentially Mediates the Formation and Maintenance of Monogamous Pair Bonds." *Nature Neuroscience*, January 2006.

Aslan, A., et al. "Penile Length and Somatometric Parameters: A Study in Healthy Young Turkish Men." *Asian Journal of Andrology*, March 2011.

Axelrod, V. "The Fusiform Face Area: In Quest of Holistic Face Processing." *Journal of Neuroscience*, June 30, 2010.

Barnhart, K., et al. "Baseline Dimension of the Human Vagina." *Human Reproduction*, February 14, 2006.

Bos., P., et al. "Acute Effects of Steroid Hormones and Neuropeptides on Human Social-Emotional Behavior: A Review of Single Administration Studies." *Frontiers in Neuroendocrinology*, January 21, 2011.

Boulvain, M., et al. "Mechanical Methods for Induction of Labor." *Cochrane Database of Systematic Reviews*, Issue 4, 2001.

Buchheim, A., et al. "Oxytocin Enhances the Experience of Attachment Security." *Psychoneuroendocrinology*, October 2009.

Burkett, J., et al. "Activation of μ-Opioid Receptors in the Dorsal Striatum is Necessary for Adult Social Attachment in Monogamous Prairie Voles." *Neuropsychopharmacology*, July 6, 2011.

Burnham, T., and B. Hare. "Engineering Human Cooperation: Does Involuntary Neural Activation Increase Public Goods Contributions?" *Human Nature*, June 2005.

Burri, A., et al. "The Acute Effects of Intranasal Oxytocin Administration on Endocrine and Sexual Function in Males." *Psychoneuroendocrinology*, June 2008.

Carmichael, M., et al. "Plasma Oxytocin Increases in the Human Sexual Response." *Journal of Clinical Endocrinology and Metabolism*, January 1987.

Damasio, A. "Brain Trust." *Nature*, June 2, 2005.

Ditzen, B. Interview with the authors, March 29, 2011.

Ditzen, B. "Intranasal Oxytocin Increases Positive Communication and Reduces Cortisol Levels During Couple Conflict." *Biological Psychiatry*, May 2009.

Ferguson, J., et al. "The Neuroendocrine Basis of Social Recognition." *Frontiers in Neuroendocrinology*, April 2002.

Ferguson, J., et al. "Social Amnesia in Mice Lacking the Oxytocin Gene." *Nature Genetics*, July 2000.

Gamer, M. "Does the Amygdala Mediate Oxytocin Effects on Socially Reinforced Learning?" *Journal of Neuroscience*, July 14, 2010.

Georgescu, M., et al. "Vaginocervical Stimulation Induces Fos in Glutamate Neurons in the Ventromedial Hypothalamus: Attenuation by Estrogen and Progesterone." *Hormones and Behavior*, October 2006.

Gonzalez-Flores, O., et al. "Facilitation of Estrous Behavior by Vaginal Cervical Stimulation in Female Rats Involves α_1-Adrenergic Receptor Activation of the Nitric Oxide Pathway." *Behavioral Brain Research*, January 25, 2007.

Goodson, J., et al. "Mesotocin and Nonapeptide Receptors Promote Estrildid Flocking Behavior." *Science*, August 14, 2009.

Griskevivius, V., et al. "Blatant Benevolence and Conspicuous Consumption: When Romantic Motives Elicit Strategic Costly Signals." *Journal of Personality and Social Psychology* 93, no. 1 (2007).

Guastella, A., et al. "Intransal Oxytocin Improves Emotion Recognition for Youth with Autism Spectrum Disorders." *Biological Psychiatry*, November 7, 2009.

Guastella, A., et al. "Oxytocin Increases Gaze to the Eye Region of Human Faces." *Biological Psychiatry*, September 21, 2007.

Heinrichs, M. Interview with the authors, March 30, 2011.

Heinrichs, M., et al. "Effects of Suckling on Hypothalamic-Pituitary-Adrenal Axis Responses to Psychosocial Stress in Postpartum Lactating Women." *Journal of Clinical Endocrinology and Metabolism*, October 2001.

Israel, S., et al. "The Oxytocin Receptor (*OXTR*) Contributes to Prosocial Fund Allocations in the Dictator Game and the Social Value Orientations Task." *PLoS One*, May 20, 2009.

Jhirad, A., and T. Vago. "Induction of Labor by Breast Stimulation." *Obstetrics and Gynecology*, March 1973.

Kavanagh, J., et al. "Sexual Intercourse for Cervical Ripening and Induction of Labor." *Cochrane Database of Systematic Reviews*, no. 1 (2007).

Kavanagh, J., et al. "Breast Stimulation for Cervical Ripening and Induction of Labour." *Cochrane Database of Systematic Reviews,* no. 4 (2001).

Kendrick, K., et al. "Importance of Vaginocervical Stimulation for the Formation of Maternal Bonding in Primiparous and Multiparous Parturient Ewes." *Physiology and Behavior,* September 1991.

Keverne, E., et al. "Vaginal Stimulation: An Important Determinant of Maternal Bonding in Sheep." *Science,* January 7, 1983.

Khan, S., et al. "Establishing a Reference Range for Penile Length in Caucasian British Men: A Prospective Study of 609 Men." *British Journal of Urology,* June 28, 2011.

King, J. Interview with the authors, August 18, 2010.

Kirsch, P., et al. "Oxytocin Modulates Neural Circuitry for Social Cognition and Fear in Humans." *Journal of Neuroscience,* December 7, 2005.

Komisaruk, B., et al. "Women's Clitoris, Vagina, and Cervix Mapped on the Sensory Cortex: fMRI Evidence." *Journal of Sexual Medicine,* July 2011.

Kosfeld, M., et al. "Oxytocin Increases Trust in Humans." *Nature,* June 2, 2005.

Lahr, J. "Mouth to Mouth." *New Yorker,* May 30, 2011.

Larrazolo-Lopez, A., et al. "Vaginocervical Stimulation Enhances Social Recognition Memory in Rats Via Oxytocin Release in the Olfactory Bulb." *Neuroscience,* March 27, 2008.

Leibenluft, E., et al. "Mothers' Neural Activation in Response to Pictures of Their Children and Other Children." *Biological Psychiatry,* August 15, 2004.

Levin, R., and C. Meston. "Nipple/Breast Stimulation and Sexual Arousal in Young Men and Women." *Journal of Sexual Medicine,* May 2006.

Liu, Y., and Z. Wang. "Nucleus Accumbens Oxytocin and Dopamine Interact to Regulate Pair Bond Formation in Female Prairie Voles." *Neuroscience* 121 (October 15, 2003): 537–44.

Long, H. W. *Sane Sex Life and Sane Sex Living: Some Things That All Sane People Ought to Know About Sex Nature and Sex Functioning; Its Place in the Economy of Life, Its Proper Training and Righteous Exercise.* New York: Eugenics Publishing Company, 1919, via Project Gutenberg.

Lynn, M. "Determinants and Consequences of Female Attractiveness and Sexiness: Realistic Tests with Restaurant Waitresses." *Archives of Sexual Behavior,* October 2009.

Meyer-Lindenberg, A., et al. "Oxytocin and Vasopressin in the Human Brain: Social Neuropeptides for Translational Medicine." *Nature Reviews Neuroscience,* September 2011.

The National Library of Ireland. "The Life and Works of William Butler Yeats." http://www.nli.ie/yeats/main.html.

Nitschke, J., et al. "Orbitofrontal Cortex Tracks Positive Mood in Mothers Viewing Pictures of Their Newborn Infants." *Neuroimage,* February 2004.

Noriuchi, M., et al. "The Functional Neuroanatomy of Maternal Love: Mother's Response to Infant's Attachment Behaviors." *Biological Psychiatry*, February 2008.

Normandin, J., and A. Murphy. "Somatic Genital Reflexes in Rats with a Nod to Humans: Anatomy, Physiology, and the Role of the Social Neuropeptides." *Hormones and Behavior*, February 19, 2011.

Perry, A., et al. "Intransal Oxytocin Modulates EEG mu/Alpha and Beta Rhythms During Perception of Biological Motion." *Psychoneuroendocrinology*, May 20, 2010.

Petrovic, P., et al. "Oxytocin Attenuates Affective Evaluations of Conditioned Faces and Amygdala Activity." *Journal of Neuroscience*, June 25, 2008.

Porter, R., et al. "Induction of Maternal Behavior in Non-Parturient Adoptive Mares." *Physiology and Behavior*, September 2002.

Rimmele, U., et al. "Oxytocin Makes a Face in Memory Familiar." *Journal of Neuroscience*, January 7, 2009.

Romeyer, A., et al. "Establishment of Maternal Bonding and Its Mediation by Vaginocervical Stimulation in Goats." *Physiology and Behavior*, February 1994.

Ross, H., and L. Young. "Oxytocin and the Neural Mechanisms Regulating Social Cognition and Affiliative Behavior." *Frontiers in Neuroendocrinology*, May 28, 2009.

Ross, H., et al. "Characterization of the Oxytocin System Regulating Affiliative Behavior in Female Prairie Voles." *Neuroscience*, May 2009.

Ross, H., et al. "Variation in Oxytocin Receptor Density in the Nucleus Accumbens Has Differential Effects on Affiliative Behaviors in Monogmaous and Polygamous Voles."*Journal of Neuroscience*, February 4, 2009.

Seltzer, L., et al. "Social Vocalizations Can Release Oxytocin in Humans." *Proceedings of the Royal Society B*, May 6, 2010.

Smith, A., et al. "Manipulation of the Oxytocin System Alters Social Behavior and Attraction in Pair-Bonding Primates, Callithrix penicillata." *Hormones and Behavior*, February 2010.

Spryropoulos, E., et al. "Size of External Genital Organs and Somatometric Parameters Among Physically Normal Men Younger Than 40 Years Old." *Urology*, September 2002.

Strathearn, L. Interview with the authors, August 31, 2011.

Taylor, M., et al. "Neural Correlates of Personally Familiar Faces: Parents, Partner and Own Faces." *Human Brain Mapping*, July 2009.

Tenore, J. "Methods for Cervical Ripening and Induction of Labor." *American Family Physician*, May 15, 2003.

Theodordou, A., et al. "Oxytocin and Social Perception: Oxytocin Increases Perceived Facial Trustworthiness and Attractiveness." *Hormones and Behavior*, June 2009.

Tucker, W. "Polygamy Chic." *American Enterprise Online*, March 21, 2006.

Unkel, C., et al. "Oxytocin Selectively Facilitates Recognition of Positive Sex and Relationship Words." *Psychological Science* 19, no. 11 (2008).

Vadney, D. "And the Two Shall Become One." Physicians for Life. http://www .physiciansforlife.org/content/view/1492/105/.

Waldheer, M., and I. Neumann. "Centrally Release Oxytocin Mediates Mating-Induced Anxiolysis in Male Rats." *Proceedings of the National Academy of Sciences*, October 16, 2007.

Walum, H., et al. "Variation in the Oxytocin Receptor Gene (*OXTR*) Is Associated with Pair-Bonding and Social Behavior." Presented at 2011 Behavior Genetics Association 41st Annual Meeting, Newport, Rhode Island, June 8, 2011.

Young, L., and X. Wang. "The Neurobiology of Pair Bonding." *Nature Neuroscience*, September 26, 2004.

Young, L., et al. "Cellular Mechanisms of Social Attachment." *Hormones and Behavior*, September 2001.

BE MY TERRITORY

Albers, H. "The Regulation of Social Recognition, Social Communication and Aggression: Vasopressin in the Social Behavior Neural Network." *Hormones and Behavior*, November 2011.

Arch, S., and P. Narins. "Sexual Hearing: The Influence of Sex Hormones on Acoustic Communication in Frogs." *Hearing Research*, June 2009.

Bos, P., et al. "Acute Effects of Steroid Hormones and Neuropeptides on Human Social-Emotional Behavior: A Review of Single Administration Studies." *Frontiers in Neuroendocrinology*, January 21, 2011.

Bos, P., et al. "Testosterone Decreases Trust in Socially Naïve Humans." *Proceedings of the National Academy of Sciences*, May 24, 2010.

Burnham, T. "High-Testosterone Men Reject Low Ultimatum Game Offers." *Proceedings of the Royal Society B*, July 5, 2007.

Carter, S., et al. "Consequences of Early Experiences and Exposure to Oxytocin and Vasopressin Are Sexually Dimorphic." *Developmental Neuroscience*, June 17, 2009.

Dewan, A., et al. "Arginine Vasotocin Neuronal Phenotypes Among Congeneric Territorial and Shoaling Reef Butterflyfishes: Species, Sex, and Reproductive Season Comparisons." *Journal of Neuroendocrinology*, December 2008.

Donaldson, Z. Interview with the authors, July 25, 2011.

Donaldson, Z., and L. Young. "Oxytocin, Vasopressin, and the Neurogenetics of Sociality." *Science*, November 6, 2008.

Donaldson, Z., et al. "Central Vasopressin V1a Receptor Activation Is Independently Necessary for Both Partner Preference Formation and Expression in Socially Monogamous Male Prairie Voles." *Behavioral Neuroscience*, February 2010.

Ebstein, R., et al. "Genetics of Human Social Behavior." *Neuron*, March 25, 2010.

Etkin, A., et al. "Emotional Processing in Anterior Cingulate and Medial Prefrontal Cortex." *Trends in Cognitive Sciences*, February 2011.

Ferguson, J., et al. "The Neuroendocrine Basis of Social Recognition." *Frontiers in Neuroendocrinology* 23 (2002): 200–224.

Fitzgerald, F. *The Great Gatsby*. New York: Scribner's, 1953.

Francis, D., et al. "Naturally Occurring Differences in Maternal Care Are Associated with the Expression of Oxytocin and Vasopressin (V1a) Receptors: Gender Differences." *Journal of Neuroendocrinology*, May 2002.

French, K. Interview with the authors, October 4, 2011.

Gobrogge, K., et al. "Anterior Hypothalamic Vasopressin Regulates Pair-Bonding and Drug-Induced Aggression in a Monogamous Rodent." *Proceedings of the National Academy of Sciences*, November 10, 2009.

Gospic, K., et al. "Limbic Justice—Amygdala Involvement in Immediate Rejection in the Ultimatum Game." *PLoS Biology*, May 3, 2011.

Griskevicius, V., et al. "Blatant Benevolence and Conspicuous Consumption: When Romantic Motives Elicit Strategic Costly Signals." *Journal of Personality and Social Psychology* 93, no. 1 (2007).

Guastella, A., et al. "Intranasal Arginine Vasopressin Enhances the Encoding of Happy and Angry Faces in Humans." *Biological Psychiatry*, June 15, 2010.

Hammock, E., and L. Young. "Microsatellite Instability Generates Diversity in Brain and Sociobehavioral Traits." *Science*, June 10, 2005.

Hartmann, E. *The Philosophy of the Unconscious: Speculative Results According to the Inductive Method of Physical Science*. Edinburgh: Ballantyne, Hanson, and Company, 1884.

Heinrichs, M. Interview with the authors, March 30, 2011.

Holmes, C., et al. "Science Review: Vasopressin and the Cardiovascular System Part 1—Receptor Physiology." *Critical Care*, June 26, 2003.

Homer. *The Odyssey*. New York: Penguin Classics, 1966.

Hopkins, W. Presentation, Workshop on the Biology of Prosocial Behavior, Emory University, October 23–24, 2011.

Hopkins, W., et al. "Polymorphic Indel Containing the RS3 Microsatellite in the 5^1 Flanking Region of the Vasopressin Receptor Gene Is Associated with Chimpanzee (Pan troglodytes) Personality." *Genes, Brains and Behaviour*, April 20, 2012.

Ishunina, T., and D. Swaab. "Vasopressin and Oxytocin Neurons of the Human Supraoptic and Paraventricular Nucleus; Size Changes in Relation to Age and Sex." *Journal of Clinical Endocrinology and Metabolism* 84, no. 12 (1999).

Ishunina, T., et al. "Activity of Vasopressinergic Neurons of the Human Supraoptic Nucleus is Age- and Sex-Dependent." *Journal of Neuroendocrinology*, April 1999.

Israel, S., et al. "Molecular Genetic Studies of the Arginine Vasopressin 1a Receptor (*AVPR1A*) and the Oxytocin Receptor (*OXTR*) in Human Behaviour: From

Autism to Altruism with Some Notes in Between." *Progress in Brain Research*, 2008.

Klieman, D., et al. "Atypical Reflexive Gaze Patterns on Emotional Faces in Autism Spectrum Disorders." *Journal of Neuroscience*, September 15, 2010.

Knafo, A., et al. "Individual Differences in Allocation of Funds in the Dictator Game Associated with Length of the Arginine Vasopressin 1a Receptor RS3 Promoter Region and Correlation Between RS3 Length and Hippocampal mRNA." *Genes, Brain and Behavior*, April 2008.

Levi, M., et al. "Deal or No Deal: Hormones and the Mergers and Acquisitions Game." *Management Science*, September 2010.

Lim, M., et al. "Enhanced Partner Preference in a Promiscuous Species by Manipulating the Expression of a Single Gene." *Nature*, June 17, 2004.

Lim, M., et al. "The Role of Vasopressin in the Genetic and Neural Regulation of Monogamy." *Journal of Neuroendocrinology*, April 2004.

Lynn, M. "Determinants and Consequences of Female Attractiveness and Sexiness: Realistic Tests with Restaurant Waitresses." *Archives of Sexual Behavior*, October 2009.

McGraw, L., and L. Young. "The Prairie Vole: An Emerging Model Organism for Understanding the Social Brain." *Trend in Neuroscience*, February 2010.

Meyer-Lindenberg, A. "Impact of Prosocial Neuropeptides on Human Brain Function." *Progress in Brain Research*, 2008.

Meyer-Lindenberg, A., et al. "Oxytocin and Vasopressin in the Human Brain: Social Neuropeptides for Translational Medicine." *Nature Reviews Neuroscience*, September 2011.

Meyer-Lindenberg, A., et al. "Genetic Variants in *AVPR1A* Linked to Autism Predict Amygdala Activation and Personality Traits in Healthy Humans." *Molecular Psychiatry*, October 2009.

Miczek, K., et al. "Escalated or Suppressed Cocaine Reward, Tegmental BDNF, and Accumbal Dopamine Caused by Episodic versus Continuous Social Stress in Rats." *Journal of Neuroscience*, July 6, 2011.

Mulcahy, S. Interview with the authors, October 28, 2011.

Neumann, I., et al. "Aggression and Anxiety: Social Context and Neurobiological Links." *Frontiers in Behavioral Neuroscience*, March 30, 2010.

Normandin, J., and A. Murphy. "Somatic Genital Reflexes in Rats with a Nod to Humans: Anatomy, Physiology, and the Role of the Social Neuropeptides." *Hormones and Behavior*, February 19, 2011.

Ophir, A. "Variation in Neural V1aR Predicts Sexual Fidelity and Space Use Among Male Prairie Voles in Semi-Natural Settings." *Proceedings of the National Academy of Sciences*, January 29, 2008.

"Prenatal Shaping of Behavior." *British Medical Journal*, April 25, 1964.

Prichard, Z., et al. "AVPR1A and OXTR Polymorphisms Are Associated with Sexual and Reproductive Behavioral Phenotypes in Humans." *Human Mutation*, November 2007.

Schmadeke, S. "Homer Glen Man Pleads Guilty to Killing Cat in Jealous Rage." *Chicago Tribune*, March 23, 2011.

Shalev, I., et al. "Vasopressin Needs an Audience: Neuropeptide Elicited Stress Responses Are Contingent Upon Perceived Social Evaluative Threats." *Hormones and Behavior*, April 30, 2011.

Shirtcliff, E., et al. "Neurobiology of Empathy and Callousness: Implications for the Development of Antisocial Behavior." *Behavioral Sciences and the Law*, August 2009.

Stribley, J., and S. Carter. "Developmental Exposure to Vasopressin Increases Aggression in Adult Prairie Voles." *Proceedings of the National Academy of Sciences*, October 26, 1999.

Takahashi, H., et al. "Dopamine D1 Receptors and Nonlinear Probability Weighting in Risky Choice." *Journal of Neuroscience*, December 8, 2010.

Thompson, R., et al. "The Effects of Vasopressin on Human Facial Responses Related to Social Communication." *Psychoneuroendocrinology*, January 2009.

Thompson, R., et al. "Sex-Specific Influences of Vasopressin on Human Social Communication." *Proceedings of the National Academy of Sciences*, May 16, 2006.

Todd, K. Interview with the authors, October 4, 2011.

Uzefovsky, F., et al. "Vasopressin Impairs Emotion Recognition in Men." *Psychoneuroendocrinology*, August 17, 2011.

Von Dawans, B. Interview with the authors, March 30, 2011.

Wagenaar, D., et al. "A Hormone-Activated Central Pattern Generator for Courtship." *Current Biology*, March 23, 2010.

Waldherr, M., and I. Neumann. "Centrally Released Oxytocin Mediates Mating-Induced Anxiolysis in Male Rats." *Proceedings of the National Academy of Sciences*, October 16, 2007.

Walum, H., et al. "Genetic Variation in the Vasopressin Receptor 1a Gene (*AVPR1A*) Associates with Pair-Bonding Behavior in Humans." *Proceedings of the National Academy of Sciences*, September 16, 2008.

Williamson, M., et al. "The Medial Preoptic Nucleus Integrates the Central Influences of Testosterone on the Paraventicular Nucleus of the Hypothalamus and Its Extended Circuitries." *Journal of Neuroscience*, September 1, 2010.

Winslow, J., et al. "A Role for Central Vasopressin in Pair Bonding in Monogamous Prairie Voles." *Nature*, October 7, 1993.

Young, L., and X. Wang. "The Neurobiology of Pair Bonding." *Nature Neuroscience*, September 26, 2004.

Young, L., et al. "Cellular Mechanisms of Social Attachment." *Hormones and Behavior*, September 2001.

Young, L., et al. "Increased Affiliative Response to Vasopressin in Mice Expressing the V1a Receptor from a Monogamous Vole." *Nature*, August 19, 1999.

Zink, C., et al. "Vasopressin Modulates Medial Prefrontal Cortex—Amygdala Circuitry During Emotion Processing in Humans." *Journal of Neuroscience*, May 19, 2010.

ADDICTED TO LOVE

Aragona, B., et al. "Nucleus Accumbens Dopamine Differentially Mediates the Formation and Maintenance of Monogamous Pair Bonds." *Nature Neuroscience*, January 2006.

Bosch, O. Interview with the authors, December 20, 2011.

Bosch, O., et al. "The CRF System Mediates Increased Passive Stress-Coping Behavior Following the Loss of a Bonded Partner in a Monogamous Rodent." *Neuropsychopharmacology*, October 15, 2008.

Buckholtz, J., et al. "Dopaminergic Network Differences in Human Impulsivity." *Science*, July 30, 2010.

Canetto, S., and D. Lester. "Love and Achievement Motives in Women's and Men's Suicide Notes." *Journal of Psychology*, September 2002.

The Deccan Herald. http://www.deccanherald.com/content/53967/love-affair-triggers-more-suicides.html.

Eastwick, P., et al. "Mispredicting Distress Following Romantic Breakup: Revealing the Time Course of the Affective Forecasting Error." *Journal of Experimental Social Psychology* 44 (2008): 800–807.

Fernando, R., et al. "Study of Suicides Reported to the Coroner in Columbo, Sri Lanka." *Medicine, Science and the Law*, January 2010.

Fisher, H., et al. "Reward, Addiction and Emotion Regulation Systems Associated with Rejection in Love." *Journal of Neurophysiology*, May 5, 2010.

Frohmader, K., et al. "Methamphetamine Acts on Subpopulations of Neurons Regulating Sexual Behavior in Male Rats." *Neuroscience*, March 31, 2010.

Goldstein, R., et al. "Decreased Prefrontal Cortical Sensitivity to Monetary Reward Is Associated with Impaired Motivation and Self-Control in Cocaine Addiction." *American Journal of Psychiatry*, January 2007.

Kelley, A., and K. Berridge. "The Neuroscience of Natural Rewards: Relevance to Addictive Drugs." *Journal of Neuroscience*, May 1, 2002.

Kiecolt-Glaser, J., et al. "Marital Quality, Marital Disruption, and Immune Function." *Psychosomatic Medicine* 49, no. 1 (January–February 1987).

Koob, G. Interview with the authors, December 21, 2011.

Koob, G. "The Role of CRF and CRF-Related Peptides in the Dark Side of Addiction." *Brain Research*, February 16, 2010.

Koob, G., and N. Volkow. "Neurocircuitry of Addiction." *Neuropsychopharmacology*, August 26, 2009.

Koob, G., and E. Zorrilla. "Neurobiological Mechanisms of Addiction: Focus on Corticotropin-Releasing Factor." *Current Opinion in Investigational Drugs*, January 2010.

Krawczyk, M., et al. "A Switch in the Neuromodulatory Effects of Dopamine in the Oval Bed Nucleus of the Stria Terminalis Associated with Cocaine Self-Administration in Rats." *Journal of Neuroscience*, June 15, 2011.

Krishnan, B., et al. "Dopamine Receptor Mechanisms Mediate Corticotropin-Releasing Factor-Induced Long-Term Potentiation in the Rat Amygdala Following Cocaine Withdrawal." *European Journal of Neuroscience*, March 2010.

Lester, D., et al. "Motives for Suicide—A Study of Australian Suicide Notes." *Crisis*, 2004.

Lester, D., et al. "Correlates of Motives for Suicide." *Psychological Reports*, October 2003.

Liu, Y., et al. "Social Bonding Decreases the Rewarding Properties of Amphetamine Through a Dopamine D1 Receptor-Mediated Mechanism." *Journal of Neuroscience*, June 1, 2011.

Liu, Y., et al. "Nucleus Accumbens Dopamine Mediates Amphetamine-Induced Impairment of Social Bonding in a Monogamous Rodent Species." *Proceedings of the National Academy of Sciences*, January 19, 2010.

Loewenstein, G. "Emotions in Economic Theory and Economic Behavior." *AEA Papers and Proceedings*, May 2000.

Lowenstein, G. "Out of Control: Visceral Influences on Behavior." *Organizational Behavior and Human Decision Processes* 65, no. 3 (1996).

Martin-Fardon, R., et al. "Role of Innate and Drug-Induced Dysregulation of Brain Stress and Arousal Systems in Addiction: Focus on Corticotropin-Releasing Factor; Nocieptin/Orphanin FQ, and Orexin/Hypocretin." *Brain Research*, February 16, 2010.

McGregor, I., et al. "From Ultrasocial to Antisocial: A Role for Oxytocin in the Acute Reinforcing Effects and Long-Term Adverse Consequences of Drug Use?" *British Journal of Pharmacology* 154 (2008): 358–68.

Murray, F. Interview with the authors, December 22, 2011.

Najib, A., et al. "Regional Brain Activity in Women Grieving a Romantic Relationship Breakup." *American Journal of Psychiatry*, December 2004.

Navarro-Zaragoza, J., et al. "Effects of Corticotropin-Releasing Factor Receptor-1 (CRF1R) Antagonists on the Brain Stress System Responses to Morphine Withdrawal." *Molecular Pharmacology*, February 16, 2010.

O'Connor, M. F., et al. "Craving Love? Enduring Grief Activates Brain's Reward Center." *Neuroimage*, August 15, 2008.

Petrovic, B., et al. "The Influence of Marital Status on Epidemiological Characteristics of Suicides in the Southeastern Part of Serbia." *Central European Journal of Public Health*, March 2009.

Pfaus, J. Interview with the authors, June 8–9, 2011.

Pine, A., et al. "Dopamine, Time, and Impulsivity in Humans." *Journal of Neuroscience*, June 30, 2010.

Pitchers, K., et al. "Neuroplasticity in the Mesolimbic System Induced by Natural Reward and Subsequent Reward Abstinence." *Biological Psychiatry*, May 1, 2010.

Pridmore, S., and Z. Majeed. "The Suicides of the Metamorphoses." *AustralAsia Psychiatry*, February 2011.

Rudnicka-Drozak, E., et al. "Psychosocial and Medical Conditions for Suicidal Behaviors Among Children and Young People in Lublin Province." *Wiadomosci Lekarskie*, supp. 1 (2002).

Rutherford, H., et al. "Disruption of Maternal Parenting Circuitry by Addictive Process: Rewiring of Reward and Stress Systems." *Frontiers in Psychiatry*, July 6, 2011.

Shalev, U., et al. "Role of CRF and Other Neuropeptides in Stress-Induced Reinstatement of Drug Seeking." *Brain Research*, February 16, 2010.

Stoessel, C., et al. "Differences and Similarities on Neuronal Activities of People Begin Happily and Unhappily in Love: A Functional Magnetic Resonance Imaging Study." *Neuropsychobiology*, May 24, 2011.

Takahashi, H., et al. "Dopamine D1 Receptors and Nonlinear Probability Weighting in Risky Choice." *Journal of Neuroscience*, December 8, 2010.

Troisi, A., et al. "Social Hedonic Capacity Is Associated with the A118G Polymorphism of the mu-Opioid Receptor Gene (*OPRM1*) in Adult Healthy Volunteers and Psychiatric Patients." *Social Neuroscience* 6 (2011): 88–97.

Urban, N., et al. "Sex Differences in Striatal Dopamine Release in Young Adults After Oral Alcohol Challenge: A PET Imaging Study with [^{11}C]Raclopride." *Biological Psychiatry*, October 15, 2010.

Van den Bergh, B., et al. "Bikinis Instigate Generalized Impatience in Intertemporal Choice." *Journal of Consumer Research*, December 5, 2007.

Wise, A., and M. Morales. "A Ventral Tegmental CRF-Glutamate-Dopamine Interaction in Addiction." *Brain Research*, February 16, 2010.

Wise, R. "Dopamine and Reward: The Anhedonia Hypothesis 30 Years On." *Neurotoxicity Research*, October 2008.

Young, K., et al. "Amphetamine Alters Behavior and Mesocorticolimbic Dopamine Receptor Expression in the Monogamous Female Prairie Vole." *Brain Research*, January 7, 2011.

Younger, J., et al. "Viewing Pictures of a Romantic Partner Reduces Experimental Pain: Involvement of Neural Reward Systems." *PLoS One*, October 13, 2010.

THE INFIDELITY PARADOX

Aragona, B., et al. "Nucleus Accumbens Dopamine Differentially Mediates the Formation and Maintenance of Monogamous Pair Bonds." *Nature Neuroscience*, January 2006.

Ashton, G. "Mismatches in Genetic Markers in a Large Family Study." *American Journal of Human Genetics* 32 (1980): 601–13.

Baugh, A., ed. *Chaucer's Major Poetry*. Upper Saddle River, NJ: Prentice-Hall, 1963.

Buckholtz, J. Interview with the authors, January 19, 2012.

Buckholtz, J., et al. "Dopaminergic Network Differences in Human Impulsivity." *Science*, July 30, 2010.

Bullough, V., and J. Brundage. *Handbook of Medieval Sexuality*. New York: Garland, 2000.

Burri, A., and T. Spector. "The Genetics of Sexual Behavior." *Behavioral Genetics*, May 2008.

Cherkas, L., et al. "Genetic Influences on Female Infidelity and Number of Sexual Partners in Humans: A Linkage and Association Study of the Role of the Vasopressin Receptor Gene (*AVPR1A*)." *Twin Research*, December 2004.

Coontz, S. Interview with the authors, December 14, 2011.

Delhey, K., et al. "Paternity Analysis Reveals Opposing Selection Pressures on Crown Coloration in the Blue Tit." *Proceedings of The Royal Society B*, October 2003.

Diekhof, E., and O. Gruber. "When Desire Collides with Reason: Functional Interactions Between Anteroventral Prefrontal Cortex and Nucleus Accumbens Underlie the Human Ability to Resist Impulsive Desires." *Journal of Neuroscience*, January 27, 2010.

Durante, K., and N. Li. "Oestradiol Level and Opportunistic Mating in Women." *Biology Letters*, April 23, 2009.

Ebstein, R., et al. "Genetics of Social Behavior." *Neuron*, March 25, 2010.

Fidler, A., et al. "DRD4 Gene Polymorphisms Are Associated with Personality Variation in a Passerine Bird." *Proceedings of The Royal Society B*, July 22, 2007.

Fiorino, D., et al. "Dynamic Changes in Nucleus Accumbens Dopamine Efflux During the Coolidge Effect in Male Rats." *Journal of Neuroscience*, June 15, 1997.

Foerster, K., et al. "Females Increase Offspring Heterozygosity and Fitness Through Extra-Pair Matings." *Nature*, October 16, 2003.

Forstmeier, W., et al. "Female Extrapair Mating Behavior Can Evolve via Indirect Selection on Males." *Proceedings of the National Academy of Sciences*, June 28, 2011.

French, J. Presentation, Workshop on the Biology of Prosocial Behavior, Emory University, October 23–24, 2011.

Gangestad, S., et al. "Women's Sexual Interests Across the Ovulatory Cycle Depend on Primary Partner Developmental Instability." *Proceedings of the Royal Society B*, August 17, 2005.

Gangestad, S., et al. "Changes in Women's Sexual Interests and Their Partners' Mate-Retention Tactics Across the Menstrual Cycle: Evidence for Shifting Conflicts of Interest." *Proceedings of the Royal Society B*, April 22, 2002.

Garcia, J., et al. "Associations Between Dopamine D4 Receptor Gene Variation with Both Infidelity and Sexual Promiscuity." *PLoS One*, November 30, 2010.

Garland, R. *The Greek Way of Life*. Ithaca, NY: Cornell University Press, 1990.

Gjedde, A., et al. "Inverted-U-Shaped Correlation Between Dopamine Receptor Availability and Striatum and Sensation Seeking." *Proceedings of the National Academy of Sciences*, February 23, 2010.

Gratian. http://www.fordham.edu/halsall/source/gratian1.asp.

Hammock, E., and L. Young. "Microsatellite Instability Generates Diversity in Brain and Sociobehavioral Traits." *Science*, June 10, 2005.

Havens, M., and J. Rose. "Investigation of Familiar and Novel Chemosensory Stimuli by Golden Hamsters: Effects of Castration and Testosterone Replacement." *Hormones and Behavior*, December 1992.

Heinrichs, M. Interview with the authors, March 30, 2011.

Hostetler, C., et al. "Neonatal Exposure to the D1 Agonist SKF38393 Inhibits Pair Bonding in the Adult Prairie Vole." *Behavioral Pharmacology*, October 2011.

Johnston, R., and K. Rasmussen. "Individual Recognition of Female Hamsters by Males: Role of Chemical Cues and of the Olfactory and Vomeronasal Systems." *Physiology and Behavior*, July 1984.

Jordan, L., and R. Brooks. "The Lifetime Costs of Increased Male Reproductive Effort: Courtship, Copulation, and the Coolidge Effect." *Journal of Evolutionary Biology*, November 2010.

Kinsey Institute for Research in Sex, Gender, and Reproduction. http://www.iub.edu/~kinsey/resources/FAQ.html#frequency.

Koene, J., and A. Ter Maat. "Coolidge Effect in Pond Snails: Male Motivation in a Simultaneous Hermaphrodite." *BMC Evolutionary Biology*, November 2007.

Korsten, P., et al. "Association Between DRD4 Gene Polymorphism and Personality Variation in Great Tits: A Test Across Four Wild Populations." *Molecular Ecology*, January 2011.

Lammer, J., et al. "Power Increases Infidelity Among Men and Women." *Psychological Science*, July 2011.

Laumann, E., et al. *The Social Organization of Sexuality*. Chicago: University of Chicago Press, 1994.

Lysias. *On the Murder of Eratosthenes*. English translation by W. R. M. Lamb. Cambridge, MA: Harvard University Press, 1930. http://www.perseus.tufts.edu/hopper/text?doc=Perseus%3Atext%3A1999.01.0154%3Aspeech%3D1.

McIntyre, M., et al. "Romantic Involvement Often Reduces Men's Testosterone Levels—But Not Always: The Moderating Role of Extrapair Sexual Interest." *Journal of Personality and Social Psychology*, October 2009.

Ophir, A., et al. "Variation in Neural V1aR Predicts Sexual Fidelity and Space Use Among Male Prairie Voles in Semi-Natural Settings." *Proceedings of the National Academy of Sciences*, January 29, 2008.

Patrick, S., et al. "Promiscuity, Paternity, and Personality in the Great Tit." *Proceedings of The Royal Society B*, November 30, 2011.

Pfaus, J. Interview with the authors, June 8–9, 2011.

Pine, A., et al. "Dopamine, Time and Impulsivity in Humans." *Journal of Neuroscience*, June 30, 2010.

Pruessner, J., et al. "Dopamine Release in Response to a Psychological Stress in Humans and Its Relationship to Early Life Maternal Care: A Positron Emission Tomography Study Using [¹¹C]Raclopride." *Journal of Neuroscience*, March 17, 2004.

Rupp, H., et al. "Partner Status Influences Women's Interest in the Opposite Sex." *Human Nature*, March 1, 2009.

Ryan, C., and C. Jethá. http://www.sexatdawn.com/page4/page4.html.

Schmitt, D., et al. "Patterns and Universals of Mate Poaching Across 53 Nations: The Effects of Sex, Culture, and Personality on Romantically Attracting Another Person's Partner." *Journal of Personality and Social Psychology*, April 2004.

Schoebi, D., et al. "Stability and Change in the First 10 Years of Marriage: Does Commitment Confer Benefits Beyond the Effects of Satisfaction?" *Journal of Personality and Social Psychology*, November 21, 2011.

Solomon, N. "Polymorphism at the avpr1a locus in male prairie voles correlated with genetic but not social monogamy in field populations." *Molecular Ecology*, November 2009.

Steiger, S., et al. "The Coolidge Effect, Individual Recognition and Selection for Distinctive Cuticular Signatures in a Burying Beetle." *Proceedings, Biological Sciences, the Royal Society B*, August 2008.

Strathearn, L. Interview with the authors, August 30, 2011.

Terris, M., and M. Oalmann. "Carcinoma of the Cervix: An Epidemiologic Study." *Journal of the American Medical Association*, December 3, 1960.

Walker, R., et al. "Evolutionary History of Partible Paternity in Lowland South America." *Proceedings of the National Academy of Sciences*, October 25, 2010.

Wilson, J., et al. "Modification in the Sexual Behavior of Male Rats Produced by Changing the Stimulus Female." *Journal of Comparative Physiology and Psychology*, June 1963.

REWRITING THE STORY OF LOVE?

Ablow, K. "Don't Let Your Kids Watch Chaz Bono on *Dancing with the Stars*." Foxnews.com, September 2, 2011.

Alexander, B. "Special Report: The New Boys' Health Scare." *Redbook*, June 2011.

American Family Association. "Obama Appointing the Mentally Diseased to Prominent Public Policy Positions." http://action.afa.net/Blogs/BlogPost.aspx?id=2147491010, January 11, 2010.

Andari, E., et al. "Promoting Social Behavior with Oxytocin in High-Functioning Autism Spectrum Disorders." *Proceedings of the National Academy of Sciences*, February 16, 2010.

Bedi, G., et al. "Is Ecstasy an 'Empathogen'?" *Biological Psychiatry*, December 15, 2010.

Brinn, L. "Brain Scans Could Be Marketing Tool of the Future." *Nature*, March 4, 2010.

Centers for Disease Control and Prevention. http://www.cdc.gov/nchs/data/databriefs/db76.htm.

Chen, F., et al. "Common Oxytocin Receptor Gene (*OXTR*) Polymorphism and Social Support Interact to Reduce Stress in Humans." *Proceedings of the National Academy of Sciences*, December 13, 2011.

Combs, A. http://www.alan.com/2010/01/04/meet-amanda-simpson-likely-the-first-presidential-transgendered-appointee/.

Dando, M. "From Nose to Brain: New Route for Chemical Incapacitation?" http://www.thebulletin.org/node/8400.

De Dreu, C., et al. "The Neuropeptide Oxytocin Regulates Parochial Altruism in Intergroup Conflict Among Humans." *Science*, June 11, 2010.

Ebstein, R., et al. "Genetics of Social Behavior." *Neuron*, March 25, 2010.

Eligon, J. "Lawsuits Challenge New York City on Sex-Change Rule." *New York Times*, March 23, 2011.

Elsea, J. "Lawfulness of Interrogation Techniques Under the Geneva Conventions." Congressional Research Service, September 8, 2004.

Evans, R. "Arms Expert Warns New Mind Drugs Eyed by Military." Reuters, August 19, 2009.

Family Research Council. "Don't Let Congress and President Obama Force American Employers to Hire Homosexuals, Transsexuals and Cross-dressers." http://www.frc.org/get.cfm?i=AL10A01&f=PG07J22, January 6, 2010.

Flew, A., ed. *Body, Mind, and Death*. New York: Macmillan, 1977.

French, K. Interview with the authors, October 4, 2011.

Golden, J. Review of *The People of Great Russia: A Psychological Study* by Geoffrey Gorer and John Rickman. *American Anthropologist* 54 (1952).

Guastella, A. Interview with the authors, October 23, 2011.

Hartmann, E. *The Philosophy of the Unconscious: Speculative Results According to the Inductive Method of Physical Science*. Edinburgh: Ballantyne, Hanson, and Company, 1884.

Heinrichs, M. Interview with the authors, March 30, 2011.

Hotchkiss, A., et al. "Fifteen Years After 'Wingspread'—Environmental Disruptors and Human and Wildlife Health: Where We Are Today and Where We Need to Go." *Toxicological Sciences*, February 16, 2008.

Israel, S., et al. "The Oxytocin Receptor (OXTR) Contributes to Prosocial Fund Allocations in the Dictator Game and the Social Value Orientations Task." *PLoS One*, May 20, 2009.

Kim, H., et al. "Culture, Distress and Oxytocin Receptor Polymorphism (OXTR) Interact to Influence Emotional Support Seeking." *Proceedings of the National Academy of Sciences*, August 19, 2010.

Kim, P., et al. "The Plasticity of Human Maternal Brain: Longitudinal Changes in Brain Anatomy During the Early Postpartum Period." *Behavioral Neuroscience* 124, no. 5 (2010).

Linakis, J., et al. "Emergency Department Visits for Injury in School-Age Children in the United States: A Comparison of Nonfatal Injuries Occurring Within and Outside of the School Environment." *Academic Emergency Medicine*, May 2006.

Lowrey, A. "Programs That Tie Funding to Effectiveness Are at Risk." *New York Times*, December 2, 2011.

Macdonald, K., and D. Feifel. "Dramatic Improvement in Sexual Function Induced by Intranasal Oxytocin." *Journal of Sexual Medicine*, March 28, 2012.

Mason, A., ed. *Free Government in the Making: Readings in American Political Thought*. New York: Oxford University Press, 1965.

Meyer-Lindenberg, A., et al. "Oxytocin and Vasopressin in the Human Brain: Social Neuropeptides for Translational Medicine." *Nature Review Neuroscience*, September 2011.

Modi, M., and L. Young. "The Oxytocin System in Drug Discovery for Autism: Animal Models and Novel Therapeutic Strategies." *Hormones and Behavior*, March 2012.

Morgan, B., et al. "Should Neonates Sleep Alone?" *Biological Psychiatry*, November 1, 2011.

Palanza, P., et al. "Effects of Developmental Exposure to Bisphenol A on Brain and Behavior in Mice." *Environmental Research*, October 2008.

Pedersen, C. "Biological Aspects of Social Bonding and the Roots of Human Violence." *Annals of the New York Academy of Science*, December 2004.

Postman, N. *Technopoly*. New York: Vintage Books, 1993.

Raytheon GLBTA News. http://ai.eecs.umich.edu/people/conway/TS/O&E/Raytheon/Raytheon%20Adds%20GI&E.html, August–October 2005.

Reshetar, J. *The Soviet Polity: Government and Politics in the U.S.S.R.* New York: Harper and Row, 1978.

Rilling, J., et al. "Effects of Intranasal Oxytocin and Vasopressin on Cooperative Behavior and Associated Brain Activity in Men." *Psychoneuroendocrinology*, August 11, 2011.

Saunders, G. "Escape from Spiderhead." *New Yorker*, December 20, 2010.

Seltzer, L., et al. "Social Vocalizations Can Release Oxytocin in Humans." *Proceedings of the Royal Society B*, May 12, 2010.

Shepard, K., et al. "Genetic, Epigenetic and Environmental Impact on Sex Differences in Social Behavior." *Physiology and Behavior*, May 2009.

Stanford Encyclopedia of Philosophy. "Prisoner's Dilemma." http://plato.stanford .edu/entries/prisoner-dilemma/, 2007.

Strathearn, L. Interview with the authors, August 30, 2011.

Tomlinson, W. Interview with Robert Heath. http://www.archive.org/details/ WallaceTomlinsonInterviewingRobertHeath_March51986.

Traditional Values Coalition. "Obama Appoints She-Male to Commerce Post." http:// www.traditionalvalues.org/read/3826/obama-appoints-shemale-to-commerce-post/, January 7, 2010.

Vero Labs. https://www.verolabs.com/Default.asp.

Vom Saal, F., et al. "Chapel Hill Bisphenol A Expert Panel Consensus Statement: Integration of Mechanisms, Effects in Animals, and Potential to Impact Human Health at Current Levels of Exposure." *Reproductive Toxicology*, August–September 2007.

Wolpe, P. R. Interview with the authors, October 24, 2011.

INDEX

Limbic system, and arousal, 74–75, 80–81, 87
Liquid Trust, 240
Lizards, appetitive behavior, 2, 38–39
Loewenstein, George, 62–63, 75–76, 79, 88
Long, H. W., Dr., 124–25, 150, 151, 217
Lordosis
    human version of, 37–38, 41, 44, 47
    and prenatal hormones, 17
    primates with attachment issues, 120
    rodents, hormonal stimulation of, 19, 44,
        69–70, 75
Loss
    gender-based effects, 203, 205–6
    of romantic bond, 198–202, 205–9
Loving v. Virginia, 257
Luteinizing hormone (LH), 38
Lysias, 213–14

Machihembra, girls becoming boys
        phenomenon, 7–8, 11–16
Mallory, Sir Thomas, 215
Marriage, sexual interest, decline in, 222–25
Marshall, Maria, 90–92, 104–6, 109–15, 119,
        122–23
Masters, William, 13–14
Mate guarding
    by male of ovulating female, 57, 232
    sociological factors, 231–32
    and vasopressin, 169–73
Maternal behavior, 90–123
    animal subjects, 93, 95–96, 98–99, 101,
        103–4, 106–10
    and breast feeding, 100–102
    and delivery method, 102–3, 247
    and dopamine, 99, 101–3, 108
    epigenetic legacy of, 106–10
    neglect and drug addiction, 203
    neural circuit for, 97–104, 108
    and oxytocin, 98–100, 107, 117–19, 122
    shutting down, experimental, 98–99
    See also Mother-infant bond
Mating, drive for. *See* Appetitive behaviors;
        Appetitive reward; Consummatory
        behavior; Consummatory reward
Meaney, Michael, 106–7, 115
Medial orbitofrontal cortex, ovulation and
        risk taking, 47–48
Medial preoptic area (MPOA)
    and appetitive behavior, 74–75
    and arousal, 74–75

and maternal behavior, 98–99, 108
    size and gender, 18–19
    and territoriality, 160
Melanocyte-stimulating hormone (MSH), 75
    therapies, future view, 245
Menstrual cycle, hormonal events of, 38
Meston, Cindy, 150–51
Methylation process, 107
Michael, Richard, 42, 48
Microsatellites, vasopressin, expression
        differences, 164–69
Miller, Geoffrey, 52–55, 71
Milner, Peter, 66, 74
Mischel, Walter, 192
Modi, Meera, 245
Money, -sex relationship, 73
Money, John, social construction of gender,
        9–11, 13–15, 256
Monogamy
    inducing in nonmonogamous animals,
        165–66
    mate guarding, 169–73, 231–32
    prairie voles, 115–16, 130–32, 162–66
    and territoriality, 160–61
    and vasopressin, 162–69
Mother-infant bond
    animal subjects, 146–48
    attachment styles, 117–19
    critical period, 108–9
    and delivery method, 102–3
    and empathy development, 111–15
    and later anxiety of child, 106–10
    and later sexual behavior, 119–21
    and mother-child social interaction, 100–102
    and mothering style of adult, 117–19
    orphans without, damage to, 91–92,
        104–6, 109–15, 119, 122–23
    romantic bond compared to, 126–27, 145,
        151–53, 202–3
    and single-parent families, 115–16
    and working mothers, 250–51
    See also Maternal behavior
Mulcahy, Sean, 170–73, 179, 226–27
Mu-opioid receptor gene, 209
Murphy, Anne, 148
Murray, Fred, 185–87, 190–96, 206–9,
        211, 225

Nasal sprays, oxytocin, 140
Natural selection, 162